Atlas of Cyberspace

Atlas of Cyberspace

Martin Dodge and Rob Kitchin

Harlow, England ■ London ■ New York ■ Reading, Massachusetts ■ San Francisco ■
Toronto ■ Don Mills, Ontario ■ Sydney ■ Tokyo ■ Singapore ■ Hong Kong ■ Seoul ■
Taipei ■ Cape Town ■ Madrid ■ Mexico City ■ Amsterdam ■ Munich ■ Paris ■ Milan ■

ADDISON-WESLEY
an imprint of
Pearson Education

PEARSON EDUCATION LIMITED

Head Office:
Edinburgh Gate
Harlow CM20 2JE
Tel: +44 (0)1279 623623
Fax: +44 (0)1279 431059

London Office:
128 Long Acre
London WC2E 9AN
Tel: +44 (0)20 7447 2000
Fax: +44 (0)20 7240 5771
Website: www.it-minds.com

First published in Great Britain in 2001

ISBN 0-201-74575-5

British Library Cataloguing in Publication Data
A CIP catalogue record for this book can be obtained from the British Library.

Library of Congress Cataloging in Publication Data
Applied for.

10 9 8 7 6 5 4 3 2 1

Designed by Sue Lamble
Typeset by Pantek Art Ltd, Maidstone, Kent
Printed and bound in Italy

The Publishers' policy is to use paper manufactured from sustainable forests.

Contents

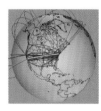

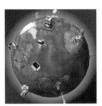

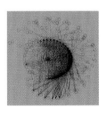

Preface

It is now over 30 years since the first Internet connection was made, between nodes installed at UCLA and Stanford University in the United States. Since then, a vast network of information and communications infrastructure has encircled the globe supporting a variety of cyberspace media – email, chat, the Web, and virtual worlds. Such has been the rapid growth of these new communications methods that by the end of 2000 there were over 400 million users connected to the Internet.

Accompanying this growth in the infrastructure, the numbers of users and the available media has been the formation of a new focus for cartography: mapping cyberspace. Maps have been created for all kinds of purposes, but the principal reasons are: to document where infrastructure is located; to market services; to manage Internet resources more effectively; to aid searching, browsing and navigating on the Web; and to explore potential new interfaces to different cyberspace media. In creating these maps, cartographers have used innovative techniques that open up new ways to understand the world around us.

This is the first book to draw together the wide range of maps produced over the last 30 years or so to provide a comprehensive atlas of cyberspace and the infrastructure that supports it. Over the next 300 or so pages, more than 100 different mapping projects are detailed, accompanied by full-colour example maps and an explanation as to how they were created.

Martin Dodge and Rob Kitchin
www-london.uk-maynooth.ie-cyberspace.net
December 2000

Acknowledgements

The *Atlas of Cyberspace* represents five years' worth of research, collating maps and research papers, and interviewing the maps' creators. In that time, many people have helped us. We are grateful to all those who assisted us in the writing and production of the *Atlas of Cyberspace*, particularly those who generously allowed us to feature maps and images of their work.

Special thanks are due to the following who went out of their way to help: Paul Adams, Keith Andrews, Richard Bartle, Mike Batty, Tim Bray, Peter Burden, Stuart Card, Chaomei Chen, Bill Cheswick, Ed Chi, K Claffy, Paul Cluskey, John Cugini, Judith Donath, Steve Eick, Gunilla Elam, Ben Fry, Joe Gurman, Muki Haklay, Nigel Hayward, Andy Hudson-Smith, Young Hyun, Jon Ippolito, Charles Lee Isbell Jr, Marty Lucas, Ernest Luk, Paul Kahn, Kate McPherson and family, Carl Malamud, Jessica Marantz, Fumio Matsumoto, Tamara Munzner, Bonnie Nardi, Marcos Novak, Linda Peake, Larry Press, Henry Ritson, Greg Roelofs, Warren Sack, Peter Salus, Gareth Smith, Marc Smith, Greg Staple, Paul Torrens, Roland Vilett, Martin Wattenberg, Darren Williams, Patrick Warfolk, Matt Zook, Mary Goodwin and Catherine Seigneret (The Cable & Wireless Archives, Porthcurno Cornwall, UK). We would also like to thank the team at Pearson – Michael Strang, Sally Carter and Katherin Ekstrom – for their enthusiastic support of this project.

Whilst every effort was made to contact copyright holders of the maps and images, we apologise for any inadvertent omissions. If any acknowledgement is missing, it would be appreciated if contact could be made (care of the publisher) so that this can be rectified in any future edition.

If you have any comments, questions or suggestions, we can be contacted at: authors@AtlasofCyberspace.com

Cover shows the Solar and Heliospheric Observatory (SOHO) Extreme ultraviolet Imaging Telescope (EIT) image. SOHO is a mission of international cooperation between ESA and NASA. http://soho.nascom.nasa.gov/

chapter 1

Mapping cyberspace

For thousands of years, people have been creating maps of the world around them – cave paintings, drawings in the sand, maps made of sticks and shells, black-and-white pencil sketches, richly colored manuscripts, three-dimensional models and, more recently, satellite images and computer-generated simulations. Since the Renaissance period, cartographers have collected together paper maps to create atlases. This book is the first comprehensive atlas of cyberspace.

Inherent in the creation of maps is the realization by the cartographer that spatial modes of communication are extremely powerful. Cartography provides a means by which to classify, represent and communicate information about areas that are too large and too complex to be seen directly. Well-designed maps are relatively easy to interpret, and they constitute concentrated databases of information about the location, shape and size of key features of a landscape and the connections between them. More recently, it has been recognized that the process of spatialization – where a spatial, map-like structure is applied to data where no inherent or obvious one exists – can provide an interpretable structure to other types of data. In essence, maps and spatializations exploit the mind's ability to more readily see complex relationships in images, providing a clear understanding of a phenomenon, reducing search time, and revealing relationships that may otherwise not have been noticed. As a consequence, they form an integral part of how we understand and explain the world.

For the past five years, we have been researching and monitoring the latest "spaces" to be mapped, namely cyberspace and its supporting infrastructure. In this book we draw together a selection of the maps and spatializations created by a range of academic and commercial "cartographers", and we examine them and the techniques used in their creation.

These maps and spatializations are extremely important for a number of reasons. First, information and communication technologies and cyberspace are having significant effects on social, cultural, political and economic aspects of everyday life.

The exact nature of these effects is contested, but evidence suggests that cyberspace is altering community relations and the bases for personal identity; is changing political and democratic structures; is instigating significant changes in urban and regional economies and patterns of employment; and is globalizing culture and information services. Maps and spatializations can help us to understand these implications by revealing the geographic extent and interrelations of the changes occurring.

Second, the extent and usage of cyberspace has grown very rapidly in the last decade. For example, there were over 1 billion publicly accessible Web pages as of January 2000 (likely to have tripled by January 2001), and the number of other media such as email, mailing lists, chat rooms, and virtual worlds has also grown significantly. Moreover, these media are used by a rapidly expanding population. For example, 377 million people were connected to the Internet by September 2000, an 87 percent increase from September 1999 (based on data from NUA, http://www.nua.ie). With so many media and users online, cyberspace has become an enormous and often confusing entity that can be difficult to monitor and navigate through. Maps and spatializations can help users, service providers and analysts comprehend the various spaces of online interaction and information, providing understanding and aiding navigation. Depending on their scale, some of the maps provide a powerful "big picture", giving people a unique sense of a space that is difficult to understand from navigation alone. As such, they have significant educational value by making often complex spaces comprehensible.

Third, the creators of these maps and spatializations are making significant contributions to the theory and practice of geographic and informational visualization in two ways. At a basic level, the research underlying the maps and spatializations is pushing the boundaries of visualization aesthetics and how we interact with data. At a more fundamental level, the research is experimenting with new ways to visualize complex data. Whilst some aspects of telecommunications infrastructure and cyberspace are relatively easy to map, such as plotting the networks of service providers onto conventional topographic maps (see chapter 2), other aspects are very difficult. This is because the spatial geometries of cyberspace are very complex, often fast-changing, and socially produced. Cyberspace offers worlds that, at first, often seem contiguous with geographic space, yet on further inspection it becomes clear that the space–time laws of physics have little meaning online. This is because space in cyberspace is purely relational. Cyberspace consists of many different media, all of which are constructions; that is, they are not natural but solely the production of their designers and, in many cases, users. They only adopt the formal qualities of geographic (Euclidean) space if explicitly programed to do so; and, indeed, many media – such as email – have severely limited spatial qualities. The inherent spaces that exist are often purely visual (with objects having no weight or mass) and their spatial fixity is uncertain (with spaces appearing and disappearing in a moment, leaving no trace of their existence). Trying to apply traditional mapping techniques to such spaces is all but impossible, because they often break two of the fundamental conventions that underlie Western cartography: first, that space is continuous and ordered; and second, that the map is not the territory but rather a representation of it. In many cases, such as maps of websites, the site becomes the map; territory and representation become one and the same.

Issues to consider when viewing images

On one level, it is possible to view and enjoy the images we present at face value. However, we think that the images are best viewed and interpreted in the light of several key issues. These issues can be expressed simply as a set of questions:

- Why was the map or spatialization created?

- Does the map or spatialization change the way we think about, and interact with, cyberspace?

- To what extent does the map or spatialization accurately reflect the data?

- Is the map or spatialization interpretable?

- How valid and reliable are the data used to construct the map or spatialization?

- Is the map or spatialization ethical?

These questions, in conjunction with the discussion below, can be used to construct a more nuanced and informed analysis of each image and technique. This type of analysis is important because to date most maps and spatializations have been produced and viewed quite uncritically.

The power of mapping

It has long been recognized that mapping is a process of creating, rather than revealing, knowledge. Throughout the process of creation, a large number of subjective – often unconscious – decisions are made about what to include and what to exclude, how the map will look, and what the map is seeking to communicate. In other words, a map is imbued with the values and judgements of the people who construct it. Moreover, they are undeniably a reflection of the culture and broader historical and political contexts in which their creators live. As such, maps are not objective, neutral artefacts but are constructed in order to provide particular impressions to their readers.

Maps, then, are situated, embodied and selective representations. Commonly, the messages are those of the powerful who pay for the maps to be drawn, and the ideological message is one of their choosing. As Mark Monmonier, in his book *How to Lie with Maps* (University of Chicago Press, 1991), comments:

In showing how to lie with maps, I want to make readers aware that maps, like speeches and paintings, are authored collections of information and are also subject to distortions arising from ignorance, greed, ideological blindness, or malice.

Spatializations of cyberspace similarly are the products of those who coded their construction algorithms. They are mappings designed for particular purposes. As such, they too are representations of power, and we should be careful to look beyond the data generated to question, in a broad sense, who the spatialization was made for, by whom, why it was produced, and what are the implications of its message and use.

Maps, then, can be a powerful means of communicating selected messages. This power can be illustrated by the extent to which they are being used to market various aspects of cyberspace enterprise. The provision of Internet services and infrastructure is a highly competitive business, dominated by large corporations, many of which operate globally. These corporations, as we illustrate in chapter 2, make significant use of maps in their marketing strategies. Indeed, the Internet marketing map is an important tool used to demonstrate the power of a company's network to potential customers. Considerable effort is invested in producing high-quality maps that present their networks in the best possible light. As such, Internet marketing maps fit into a long tradition of maps used by companies to promote their networks – be they shipping, airlines, or railroads.

When considering maps in the following chapters, one should question why the map has been presented in the way it has, and why it was produced at all.

The agency of mapping

As just noted, all maps are designed to either change or reaffirm the way we think about, and comprehend, the data presented. In many cases, maps or spatializations of cyberspace are designed to change the way we interact with cyberspace. A key question is thus to ask to what extent a mapping is successful in these aims: does a map or spatialization change the way we think about cyberspace, and do those that seek to offer new modes of interaction offer viable spatial interfaces that could replace or supplement current methods of data management and navigation? In other words, do the maps or spatializations achieve their aims, whether that be improving comprehension, providing new means of navigation or interaction, or selling a service?

A further set of questions relates to the effects if these aims are met. For example, in relation to improving interaction, if a method of spatialization qualitatively alters how we interact with media, how does this affect social relations within specific domains? It may well be the case that the process of mapping may actually change what it seeks to augment, altering the very nature of the medium involved.

Representation and distortion

Maps and spatializations are representations. They aim to represent, in a manner that is spatially consistent, some particular phenomenon. An age-old concern in cartography therefore relates to the extent to which maps adequately represent data. Maps necessarily depict a selective distortion of what they seek to portray, because they employ processes of generalization and classification. There are three principal ways in which maps can distort reality, and give rise to false interpretations: presentation; ecological fallacy; and omission. Each is discussed in turn next.

In making decisions about how data might be mapped, the cartographer has to decide how the data will be presented, considering issues such as projection, scale, classification, and graphic styles of symbols, colors, labeling and fonts. Each of these decisions can affect significantly how data is portrayed and thus interpreted. The map style dictates the choice of base data on which the phenomenon data will be plotted, and how the phenomenon data will be manipulated for presentation. Varying the projection of the base data can lead to maps that vary quite significantly in presentation. For example, the Mercator projection distorts factors such as area and shape in order to allow all rhumbs (lines of constant bearing) to appear as straight lines. While a map drawn in this way suggests that Greenland is approximately the same size as Africa, in reality Greenland would fit inside Africa several times.

Data of interest might be displayed individually or aggregated into units. Aggregation can create a whole set of problems. For example, how the aggregation classes are selected can lead to maps that look quite different. Moreover, the same data mapped onto differing sets of spatial units (e.g., wards, districts, counties, states) can produce significantly different spatial patterns. This is known as the Modifiable Areal Unit Problem (MAUP), which consists of two components: a scale problem and a zoning problem. MAUP problems arise because there is an assumption that we can delineate the boundaries between zones in a precise and meaningful manner, so that the area within a zone is uniform in relation to the data. Of course, this is not in fact the case, because natural spatial variation leads to gradual change across space. The difference between reality and the model can then lead to erroneous interpretation. This is known as the "ecological fallacy". Here, the aggregate characteristics of a whole population are inappropriately ascribed to individuals within populations, and the problem is commonly associated with mapping methods used to map the geography of Internet diffusion (see chapter 2).

Ecological fallacies are often the product of having to map data collected at particular territorial scales. Because the data have no subscale variability there is little choice but to map them at the scale collected. Many of the maps of the Internet are constructed using "off-the-shelf" data that are readily available for country-level aggregation. For example, in many studies of Internet diffusion and "digital divides", the same data sources – such as the World Bank, OECD, International Telecommunications Union, CIA world database and Network Wizards Internet data – are used repeatedly. These organizations publish orderly tables of statistics at the national level that can be turned into maps with ease and little thought. If there is no commentary in the analysis warning of the possible dangers of ecological fallacies, then the people who use the research data can easily be misinformed.

In many ways, national-level data collection is a logical unit choice as there is no doubt that individual experiences and institutional decisions are shaped by national-level power structures through government legislation, deregulation and subsidies. In some respects, however, it seems illogical to create maps that demarcate the Internet into the straightjacket of

national borders, especially when the data displayed (e.g., infrastructure owned and operated by global corporations) have little relationship to nation-states. The network technologies of cyberspace are forging connections and virtual groups that potentially subvert the primacy of national boundaries. These borders are relatively meaningless to logical connections and data flows that operate on a global scale. The question in these cases is therefore: "How much sense do existing political borders of the material world make when mapping cyberspace?"

The final way that maps can create false impressions is through omission. For example, many maps of infrastructure and cyberspace focus their attention – either deliberately or unconsciously – on the developed world in the West, especially the United States (and the majority of examples in this book are created by researchers and companies located there). This focus all too easily relegates other parts of the world, such as Africa, metaphorically – and sometimes literally – to the edge of the map. Pushing countries to the periphery reinforces, visually at least, the existing world hegemony in relation to the Internet. The lack of representation of the "unwired" masses on many of the maps is a particular concern. In reality, many of these countries are key to the sustenance of the information economy, providing sites of low-paid, low-skilled office work and the manufacture of computer and telecommunication components that are almost exclusively exported. Moreover, many of the most talented people in the field, such as computer programmers, are being drawn to high-tech centers such as Silicon Valley in the United States from countries such as India.

The issues outlined above affect all maps and spatializations, and yet they have been little considered so far in the mapping of infrastructure and cyberspace (although see our book *Mapping Cyberspace* (Routledge, 2000)). Although map makers can draw on solutions from generations of cartographic theory and practice in order to try to produce better representations of the data, much more consideration needs to be given to spatializations of cyberspace. Here, there are no standards by which to judge factors such as accuracy, precision, verisimilitude, mimesis and fallacy. Indeed, when data and mapping become synonymous, how do issues of representation apply? In this latter case, cyberspace may become meaningless outside its own representation. The need for standards to be set and for issues of representation to be addressed is then of paramount importance.

Level of user knowledge

As the work of cognitive cartographers over the past two decades has amply illustrated, whilst maps are effective at condensing and revealing complex relations, they are themselves sophisticated models. It is now widely recognized that maps are not "transparent" but are complex models of spatial information that require individuals to possess specific skills to understand and use them. Using a map means being able to read a map, which requires a distinct set of skills that must be learnt. This implies that a novice will learn little from a professionally produced map unless he or she knows how the map represents an area. This also applies to maps of cyberspace, particularly in the case of three-dimensional interactive spatializations, which may increase confusion and disorientation rather than reduce it.

Care needs to be exercised in relation to the design of maps, so that the target audience can understand and use the information portrayed. As far as we are aware, whilst there has been some work on the legibility and design of visual virtual worlds and hypertext, there has been little or no work on the legibility of maps of infrastructure or spatializations of cyberspace. Many of the maps we present in the following chapters are difficult to interpret without reference to the explanation in the text. The need for such reference points to the fact that the maps hold poor communicative properties, which need to be improved. Having said this, it must be recognized that many of the maps and spatializations have not been produced for a general audience, having been created as tools to aid specialist analysts in their work.

Data quality and availability

Maps and spatializations are only as accurate as the data used to underpin the representation. Therefore a key issue for those seeking to construct maps of infrastructure and spatializations of cyberspace is access to timely, accurate and representative data. Such access has always been a concern of cartographers, particularly since the Renaissance, but it has become a major issue since the widespread adoption of computer-based cartography in the form of geographic information systems in the 1980s. In particular, spatial data users are concerned about issues such as data coverage, completeness, standardization, accuracy and precision. Here, "accuracy" refers to the relationship between a measurement and its reality, and "precision" refers to the degree of detail in the reporting of a measurement. It is generally recognized that all spatial data are of limited accuracy due to inherent error in data generation (e.g., surveying) or source materials.

No standards of accuracy exist for data concerning cyberspace, and what sources there are are limited and fragmented, with no definitive or comprehensive databases. Consequently, maps can be fascinating but at the same time limited in scope, coverage and currency when compared with the wealth of statistics gathered and mapped for geographic space by government agencies such as the USGS, Ordnance Survey, and national census bureaux. This is compounded by the fact that both infrastructure and cyberspace lack central planning and a controlling authority that monitors and gathers statistics on their operation and use. In addition, the provision of both infrastructure and content services has become an intensely competitive and profitable business. As such, corporations are wary of giving away details that may aid competitors or threaten security.

Given the fast-growing and dynamic nature of both infrastructure and cyberspace, the issue of data quality and coverage is of critical importance. We are in little doubt that maps will become increasingly important for understanding the implications of cyberspace and in comprehending and navigating through cyberspace, but without suitable high-quality and up-to-date data to underpin their construction they will be of limited use. A valuable exercise is to apply the following questions to the data used to construct maps of cyberspace (adapted from *The Geographer's Craft Project* by Ken Foote and Donald Huebner):

- What is the age of the dataset?

- Where did the data come from?

- How accurate are positional and attribute features?

- Do the data seem logical and consistent?

- In what format are the data kept?

- How were the data checked?

- Why were the data compiled?

- What is the reliability of the data provider?

<http://www.colorado.edu/geography/gcraft/contents.html>

Ethics

One final issue to consider relates to the ethics and responsibility of researchers producing maps of cyberspace. As sociologist Marc Smith has argued, these new forms of maps and spatializations open up cyberspace to a new kind of surveillance, revealing interactions that were previously hidden in unused log files and databases.

The act of mapping itself may constitute an invasion of privacy. If the appeal of some media is their anonymity, then users may object to them being placed under wider scrutiny, even if individuals are unidentifiable. Here, public analysis may well represent an infringement of personal rights, especially if the individuals were not consulted beforehand. In some senses, these maps may work to shift the spaces they map from what their users consider semi-private spaces to public spaces, and thus the maps may actually change the nature of cyberspace itself. For example, how does the use of Chat Circles (see

Pages 174–5) alter the nature of social interaction within chat rooms? Here, it is important to consider the ways and the extent to which maps of cyberspace are "responsible artefacts" (i.e. ones that do not destroy what they seek to represent or enhance).

Structure of the book

Although still a relatively young field of interest, there have been literally thousands of maps and spatializations of cyberspace created to date. In the course of constructing this particular atlas, we have had to make numerous subjective decisions about which examples to include. At times, this has been a difficult process. Our strategy has been twofold: first, to include a very broad range of images and techniques that visualize as many different aspects of cyberspace and its underlying infrastructure as possible; second, to select those techniques that seem particularly innovative, in terms of both methodology and design, and that seem to offer promising avenues for further development. This inevitably means that the atlas is a partial record of attempts to visualize and spatialize cyberspace, yet at the same time it is intended to provide a balanced overview of the field.

In order to provide a coherent structure to the rest of the book, we have divided the remaining text into five chapters. Within each of the first four of these chapters, we provide a summary overview of some of the main arguments about the particular aspect of cyberspace being mapped, and a discussion of the merits, aims and uses of the maps and spatializations presented. The last chapter contains some final thoughts on the subject.

In chapter 2, we focus our attention on the interesting intersection of cyberspace and geographic space. Here, we present maps of the infrastructure that supports cyberspace, the demographics of cyberspace users, and the flow of data traffic across different scales from the local to the global. The examples discussed predominantly map the data from which they are constructed onto familiar geographic frameworks, although a few use a more abstract approach. These maps provide important insights into who owns and controls the supporting infrastructure, who has access to cyberspace, how the system can be surveyed, and how and from where cyberspace is being used. Often they are most useful for public understanding because a familiar template of real-world geography is used.

In chapter 3, we examine some fascinating ways to spatialize the Web in order to create information spaces that are comprehensible and, in some cases, navigable. We present a wide range of spatializations that have employed a variety of graphical techniques and visual metaphors so as to provide striking and powerful images that extend from two-dimensional "maps" to three-dimensional immersive landscapes. These spatializations are important because they provide interpretable images for data that were previously very difficult to understand. For example, topological structure data of traffic in the logs of a large website are almost impossible for humans to interpret, because they are held in large textual tables, tens of thousands of lines long, that provide no tangible referents other than attribute codes but that, once spatialized appropriately, are relatively easy to interpret.

Spatializations that seek to chart aspects of community and conversation are the focus of chapter 4. The primary attraction of cyberspace is its ability to foster communication between people through a variety of asynchronous (participants communicating at different times) and synchronous (participants present at the same time) media such as email, mailing lists, bulletin boards, MUDs (multi-user domains – see chapter 4), and virtual worlds. Here, we document novel ways to spatialize all these media. Although somewhat variable in their success, these spatializations are important because they seek to enrich the mode of interaction, and thus the success and pleasure of communication between users. Whilst none of the

spatializations we present has significantly altered how people currently use these media, they hold great potential to do so.

In chapter 5 we turn our attention away from geographic and informational visualization to consider the other ways in which cyberspace has been imagined, described and drawn. Here, we focus mainly on the work of artists, film makers and writers, who have been seeking to answer the question "What does cyberspace look like?". These visualizations are important, because they often provide the inspiration for the designers and creators of maps and spatializations discussed in chapter 4. As we have argued elsewhere, the influence of these artists, film makers and writers should not be underestimated. This is because they provide a popular imaginal sphere in which to question and explore the space–time configuration of cyberspace. Also, they have aesthetic and artistic worth in and of themselves, and as such they represent both the art and the science of mapping cyberspace.

Chapter 6 comprises our final thoughts for the book.

Concluding comment

There are clearly many issues to think about when viewing the maps and spatializations we present. However, although many are imperfect (to varying degrees), they are all fascinating examples of the innovative ways in which cyberspace is being mapped and spatialized. The examples we document are perhaps equivalent in stature to the real-world maps created at the start of the Renaissance period that formed the bedrock of modern cartography. The broad array of maps and spatializations we detail in the following chapters are the beginnings of what we are sure is going to be a vibrant area of research with many practical applications.

chapter 2

Mapping infrastructure and traffic

In this chapter, we provide an analysis of a range of maps designed to communicate information about the infrastructure that supports cyberspace, the demographics of users, and the type, flow and paths of data between locales and within media. These maps are by far the most commonly produced maps relating to cyberspace and, as we discuss below, are important for a variety of reasons, not least of which is their commercial and political value. Indeed, the maps we look at have all been created to either market the services or products of large corporations, or to aid our understanding of the structure, organization, operation, and use of information and communication technologies and cyberspace. They represent only a fraction of those so far produced and have been chosen primarily to demonstrate the range of different types of maps being constructed and the techniques used in their construction. A secondary consideration was their aesthetic appeal.

All the maps we consider are in the public domain and the majority are freely available to browse through on the Web. Clearly, there are many more maps that are deemed confidential to the companies and organizations that own and operate the infrastructure, either containing sensitive information or being of a practical nature primarily used by network engineers.

In the first part of this chapter, we focus our attention on maps that seek to delineate the vast array of telecommunication and computing infrastructures that support cyberspace. These infrastructures have been developed over several decades and built at the cost of many billions of dollars. They are often taken for granted because, unlike roads or railways, they are often invisible: buried underground, snaking across ocean floors, hidden inside wall conduits, or floating unseen in orbit above us. Given its invisibility, it is easy to assume that the infrastructure of cyberspace is as ethereal and virtual as the information and communication that it supports. However, the infrastructure has a physical presence that can be mapped onto geographic space (planemetrically or topologically) or displayed using techniques of spatialization.

Maps thus provide one of the best means of making sense of the vast and complex infrastructure of information and communication technologies, allowing us not only to see where those technologies are located and how they interconnect to provide the platform for cyberspace, but also to assess the social and economic implications of their distribution. As such, they reveal insights into the structures of the material (and, in turn, immaterial) aspects of cyberspace in terms of who controls and owns the systems, and how systems can be presented, marketed and surveyed.

Moreover, they reveal important information about physical access to cyberspace, because they display the complex and uneven geography of infrastructure across the world. Indeed, it is important to realize that the location and topology of infrastructure are key determinants in access to cyberspace, affecting cost, speed, reliability and the ability to connect. These maps illustrate that, on a global scale, infrastructure is concentrated in certain countries (such as the United States, United Kingdom and Scandinavia), on the national level it is concentrated in certain regions (e.g. Silicon Valley), and even in high-tech cities like San Francisco or New York there is very localized clustering. Accessing cyberspace is fragmented along traditional spatial and social divisions, with infrastructure density and variety being closely related to areas of wealth.

In the latter part of the chapter, we present maps that detail the types, flows and paths of data through and within domains in relation to geographic space. These maps reveal what volume of data is flowing through specific vectors, at what speed, and the different types of data traversing the Internet. As such, they display detailed pictures of the relationships between different locations, how well a system is performing, and what the Internet is being used for. They also reveal another important aspect of access, namely bandwidth. It is nowadays often stated that if the value of real estate is dependent upon location, then the value of a network connection is determined by bandwidth (see, for example, William J. Mitchell's book *City of Bits* (MIT Press, 1995)). Accessibility becomes redefined so that the "friction of distance" is replaced by the "bondage of bandwidth". Low bandwidth means longer connection times and thus higher costs. At present, high bandwidth is largely confined to information hotspots, mainly focussed around key universities, high-capacity data sources (e.g. telecommunications companies), and localized centers such as telecenters in rural areas.

We hope that the selection of maps that follows will provide useful insights into the infrastructure that supports cyberspace, who is using cyberspace, and how data travel through the networks. Whilst many are visually striking and persuasive, we would like to remind you to consider some of the issues raised in chapter 1. All the maps presented have been created by people with a variety of motivations and agendas. Furthermore, all the maps are selective and subjective presentations of their underlying data. No one map, then, is a "true" map of the infrastructure of cyberspace – and no such map can be created. Perhaps, even, our knowledge is diminishing as the scale and complexity of infrastructure grows and information about it becomes less open to scrutiny. It is important to interrogate the maps using the questions outlined on page 7.

Historical maps of telecommunications

Cyberspace's history is not confined to the recent past. It has a long antecedence with its roots in the development of the telegraph and telephone in the 19th century. These technologies were the first to connect distant places in order to allow the instant communication of data. In his book entitled *The Victorian Internet* (Weidenfeld & Nicolson, 1998) Tom Standage argues persuasively that all the advances in telecommunications since the telegraph have really been incremental improvements rather than revolutionary breakthroughs. We begin by presenting two maps that chart the geography of the telegraph and telephone networks at different times in the past.

The telegraph was the first technology that allowed a message to be passed between two distant places virtually instantaneously. Following Samuel Morse's famous demonstration of the first practical system in May 1844, connecting Baltimore and Washington DC, the telegraph network quickly spread across the United States. The map shows the extent of the telegraph infrastructure barely ten years later, in 1853. This detailed map was produced by Charles. B. Barr and charts the number and geographical reach of telegraph stations covering the eastern portion of North America, stretching from Quebec in the north to New Orleans in the south, and from Philadelphia on the eastern seaboard to Kansas city in the Mid-West. The map states that it is "compiled from reliable sources", although it is not clear what these are.

In terms of cartographic design, the geographic locations of the telegraph stations are represented by small black dots, labeled with the town's name (see the enlarged section of Pennsylvania). The telegraph wires linking stations into the nation wide network are shown by the thin black lines that have been generalized and simplified into straight-line segments. An underlying base map shows the coastline, state borders and major rivers and lakes, in order to provide a necessary context for the reader. Overall, the map provides a simple but effective way of showing the geography of a large and complex network topology. The arc–node method of representation is a design that many subsequent maps of network infrastructure have employed, as illustrated throughout this chapter.

The map also provides detailed and useful information about the "Tariff of Rates on the National Telegraph Lines" in the large table on the left-hand side. This lists some 670 telegraph stations in alphabetical order, from Alexandria, VA, to Zanesville, OH, and the cost of sending a message to that location from Pittsburgh (where the cartographer Charles B. Barr was based). For instance, it cost 90 cents for the first ten words of a message to be sent to Boston, and then 7 cents for every additional word.

2.1: Telegraph stations in the United States, the Canadas and Nova Scotia

chief cartographer: Charles B. Barr (Pittsburgh, PA).
aim: to map the location of telegraph stations and their connections. Table provides telegraph tariffs from Pittsburgh to all other stations.
form: telegraph system represented as an arc–node network on a simple, geographic, base map. All telegraph stations shown by a black dot and labeled with name.
technique: color paper map, 58 by 64cm, "compiled from reliable sources".
date: published in 1853.
further information: Library of Congress, Geography and Map Division, Washington DC <http://hdl.loc.gov/loc.gmd/g3701p.ct000084>
further reading: *The Victorian Internet: The Remarkable Story of the Telegraph and the Nineteenth Century's Online Pioneers*, by Tom Standage (Weidenfeld & Nicolson, 1998).

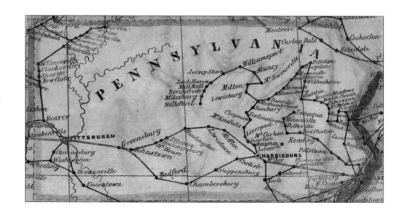

Created almost a hundred years later in 1945, the map opposite is a colorful and attractive marketing map designed to promote the global telecommunications network of Cable & Wireless. At this time, Britain was still the heart of a diminishing global empire, and C&W provided communications with the many distant colonies, colored red on the map. The map was created by MacDonald Gill, a noted artist and calligrapher in the 1930s and 1940s, whose work was published in many books and as posters. He is perhaps best known for the large murals he painted for the ocean liners *Queen Mary* and *Queen Elizabeth* of that period.

To illustrate Britain's role as the hub of this empire, a polar projection has been employed, placing Britain at the center of the map. The C&W telecommunications network consisted of a vast system of undersea cables and wireless stations that traversed the globe, and these are represented on the map by solid and dotted lines. The map is clearly designed for promotional purposes, with its rich decorative border aimed to appeal to the general public. This border consists of pictorial scenes that represent the different types of technologies employed by the company, such as wireless transmitter masts and cable-laying ships. The map, then, is a marketing map, designed to show the geographical reach of the network and the range of different services that the company employs to ensure a fast and reliable service. More recent examples of marketing maps are presented later in the chapter (see page 30).

2.2: Cable & Wireless "Great Circle" map

chief cartographer: MacDonald Gill (Cable & Wireless).
aim: to provide a marketing map showing the global connectivity of Cable & Wireless (C&W) through its telecommunications network, with Britain centered representing its position as "hub of the world".
form: an attractive map of the world with arcs representing globe-spanning C&W telecommunications network. Surrounded by rich pictorial embellishments showing telecommunication scenes.
technique: geographic map using a polar projection centered on the UK.
date: 1945.
further information: Cable & Wireless – A History <http://www.cwhistory.com/>
further reading: *America Calling: A Social History of the Telephone to 1940*, by Claude S. Fischer (University of California Press, 1994).

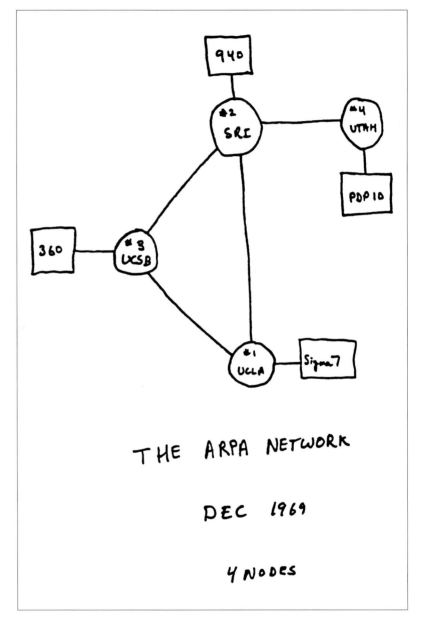

Maps from the birth of the Net

ARPANET pioneered wide-area computer networking and laid the foundations of the Internet as we know it today, developing both the technical and the social infrastructure of Internet working. It was conceived as a method to link several incompatible computer systems located at various points across the United States so that resources could be shared. (It is a popular misconception that the linking of computers was to ensure continuation of the network after a nuclear attack. The system did, however, use the idea with the intention of overcoming such an event, as proposed by RAND researcher Paul Baran in the 1960s.) It was funded by the US military, through the ARPA agency, and it was designed and operated by the Bolt Beranek and Newman (BBN) company. The first network node was installed at the University of California, Los Angeles (UCLA) in September 1969. The network was built as a distributed, decentralized system with each node of equal importance, and it used packet-switching protocols that allowed data to travel by any available route between nodes.

Two of the earliest surviving maps of the nascent ARPANET are shown opposite. The first diagram displays the first node of the network (#1 IMP) connected to a mainframe computer (#1 Host, a Sigma 7 model) at UCLA in 1969. This very simple conceptual diagram is a significant record as it marks an important moment in history, the connection of the mainframe to a message translator forming the first location in cyberspace. Test messages were passed between the network node (IMP) and the host computer on 2 September 1969. The "map" is a hand-drawn black-and-white sketch, reminiscent of "back-of-the-envelope" wiring diagrams drawn by many an engineer.

A test configuration of ARPANET's first four operational nodes as at the end of 1969 is represented by the second figure. The nodes of the network are again represented by circles, which are numbered by the order in which they were installed and labeled with the site name. So, after UCLA there was SRI (the Stanford Research Institute), then UCSB (University of California Santa Barbara), and then the University of Utah. The connections between nodes, running over special telephone lines, are shown by the straight lines. The square boxes on the map represent the actual computers connected to the network, and they are labeled with the model name – for example, PDP10 (made by DEC) and 360 (from IBM).

ARPANET grew rapidly from these initial nodes, and its expansion was charted in a fascinating series of maps used by the BBN engineers who built and managed the network. The topology of the network was plotted using both geographic

2.3: Sketch maps of ARPANET in September and December 1969

chief cartographer: unknown ARPANET scientist/engineer.
aim: to record the initial topological structure of ARPANET.
form: black-and-white line drawing of arc and nodes.
technique: hand-drawn, "back-of-envelope" style of sketches.
dates: September and December 1969.
further information: The Computer Museum History Center
<http://www.computerhistory.org/timeline/topics/networks.page>

and non-geographic layouts. The top spatialization opposite displays the ARPANET network topology as it had developed by 1977, using an abstract layout. By the time this map was drawn, ARPANET was no longer a curious experiment but a fully operational network of nearly 60 nodes. The network was run by the Defense Communication Agency and had nodes in many US states, including Hawaii, and even stretched to Europe with a link to a NATO radar facility in Norway and then on to London. These overseas links were carried by satellite circuits, which are represented by the wavy lines on the map.

The bottom map displays the same data from just a few months later, in June 1977, plotted onto a geographic base map. The most striking feature of this map is the great concentration of sites in California and the north-east of the United States, with only a scattering of nodes outside these areas. This pattern clearly reveals the density of military establishment and defense-funded, advanced-research labs in these regions. The cartographic style is simple and functional, using an arc–node representation to show the logical structure of the network. Different types of nodes are shown by circle, square and black triangle symbols drawn on the map in the approximate geographic location of the site, along with a name label. The network connections between sites are represented by black lines. Magnifying circles are employed to display the densest concentration of nodes, in northern and southern California, and the Boston and Washington regions. The paths taken by the actual connecting wires (leased from AT&T) are abstracted to straight lines for ease of representation.

Although these maps are relatively simple black-and-white line drawings, they are now of historical importance because they were created by the people who built the first Internet network. As a collection, they chart its size and approximate geographic structure. Moreover, they record what no longer exists. ARPANET has long since been decommissioned, having been officially "turned off" in 1989. It was superseded and replaced by faster, more sophisticated networks.

2.3: ARPANET logical map, March 1977

chief cartographer: unknown graphic designer (Bolt Beranek and Newman).
aim: to display the topology of connections of ARPANET in early 1977, with particular reference to host computers.
form: similar to a wiring diagram of an electrical circuit.
technique: black-and-white line drawing on paper.
date: March 1977.

2.3: ARPANET geographic map, June 1977

chief cartographer: unknown graphic designer (Bolt Beranek and Newman).
aim: to display the geographic topology of ARPANET in the summer of 1977.
form: outline map of the United States with the network represented using an arc–node technique.
technique: black-and-white line drawing on paper.
date: June 1977.
further reading: *Where Wizards Stay up Late: The Origins of the Internet*, by Katie Hafner and Matthew Lyons (Simon and Schuster, 1996). *Inventing the Internet*, by Janet Abbate (MIT Press, 1999). *Casting the Net: From Arpanet to Internet and beyond...*, by Peter H. Salus (Addison-Wesley, 1995).

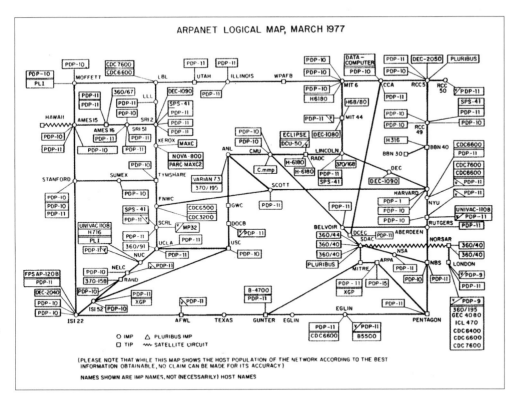

ARPANET LOGICAL MAP, MARCH 1977

(PLEASE NOTE THAT WHILE THIS MAP SHOWS THE HOST POPULATION OF THE NETWORK ACCORDING TO THE BEST INFORMATION OBTAINABLE, NO CLAIM CAN BE MADE FOR ITS ACCURACY)

NAMES SHOWN ARE IMP NAMES, NOT (NECESSARILY) HOST NAMES

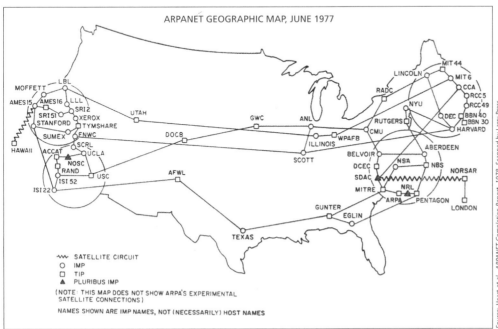

ARPANET GEOGRAPHIC MAP, JUNE 1977

(NOTE: THIS MAP DOES NOT SHOW ARPA'S EXPERIMENTAL SATELLITE CONNECTIONS)

NAMES SHOWN ARE IMP NAMES, NOT (NECESSARILY) HOST NAMES

Source: Heart et al., ARPANET Completion Report, 1978. Scanned by Larry Press

Mapping where the wires, fiber-optic cables and satellites really are

The maps we have so far presented use a topological mapping technique. That is, they provide a generalized representation of a network, showing the correct linkages between nodes but not the exact route of the connections in relation to geographic space. In this section we present three maps that seek to provide more exact representations of the actual locations of network infrastructure. The first two chart the paths of cables at different spatial scales: a building and a continent; the third map plots the path of the Teledesic satellite constellation.

In general, these maps are produced solely for use by the engineers who maintain the facilities, allowing them accurately to locate the infrastructure in an environment. For example, if an engineer needs to find a fiber-optic cable that runs under a street, an accurate map is needed to know exactly where to dig. As such, these types of maps are rarely designed for general public consumption.

Most network architecture in the built environment is invisible, running unseen in floor conduits, roof voids, underground pipes, and so on. We generally only see the connection points into which we plug our computers. The plates opposite show schematic diagrams of the network infrastructure for a floor in a building. This type of map is most often at an architectural scale and is generally very accurate, with the position of cables, network ports and cabinets plotted to the nearest centimeter.

The examples shown are from the Network Connectivity Section system which manages the large and complex network infrastructure at University College London. This network is one of the largest in the United Kingdom, serving many thousands of academics and students. The current "CRIMP" cable-management system holds a database and associated plan layouts showing the locations of over 26,000 data outlets and 331 network cabinets, spread around 118 separate buildings. With this kind of scale of network complexity, it is vital to have real-time locational information to identify and fix faults, as well as to plan for extensions and upgrades of the network. The top plate shows one floor of a building, with data ports (red diamonds) and network cabinets represented. The bottom plate is a CAD schematic of the logical cabling layout for a building.

2.4: Schematic diagram of a building's network facilities

chief cartographer: Nigel Hayward and colleagues (Network Connectivity Section, University College London).
aim: to provide a detailed layout of the network infrastructure through a building for use by network maintenance engineers.
form: computer-aided design (CAD) diagram of network topology plotted on building architecture plans.
technique: produced by a specialized cable management system called CRIMP by Cablesoft.
date: December 2000.
further information: details on CRIMP at <http://www.itracs.com/>

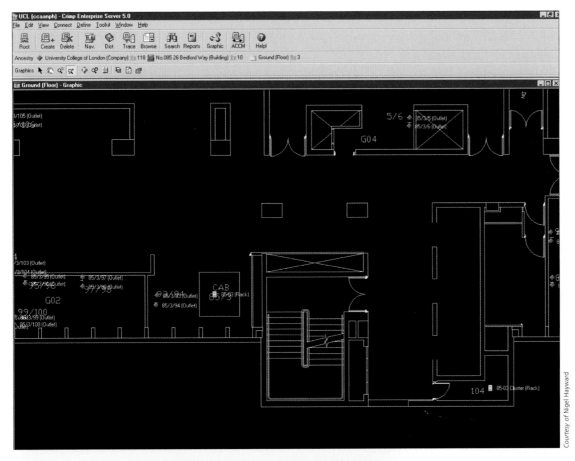

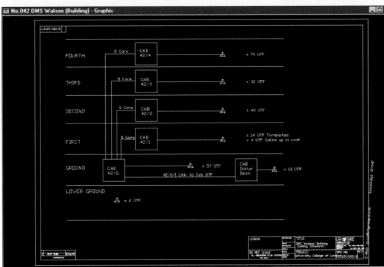

The two maps opposite are of the long undersea cables that provide vital intercontinental communications infrastructure. Undersea communication cables linking continents have been laid since the 1860s. In the 1990s their capacity was greatly increased by advances in fiber-optic technologies. A single strand of fiber-optic cable can carry great volumes of data – for example the latest cable across the North Atlantic, laid in 2000 and called TAT-14, can carry the equivalent of 9.7 million simultaneous telephone conversations. The cables consist of large bundles of fiber-optic threads housed within a protective steel casing, and can thus carry huge volumes of data across the world at the speed of light. This has led to a rapid growth in aggregate communication capacity between continents, most especially across the North Atlantic where cables connect the highly wired regions of North America and Western Europe. This means that in several regions of the world a dense network of cables criss-crosses the oceans.

The maps are produced by Alcatel Submarine Systems, a major manufacturer of telecommunications systems, and its submarine division is one of the world leaders in designing and building undersea cable systems. These maps are from a series created to show major undersea cables in different regions of the world. They are good examples of maps – aimed primarily at a technical, engineering audience – that detail the location of cable infrastructure and, most importantly, where their landing points are. These landing points are critical nodes in the system shown, for it is here that the submarine cables connect to conventional terrestrial telecom grids.

Clearly, at the scale that the maps are drawn, there is some generalization over the exact route of cables on the ocean floor. In its simple no-frills cartographic design, the map has much in common with the telegraph station map (page 12). It also uses the same kind of graphic representation – arcs and nodes – which are each labeled with their name and bandwidth.

Satellites are significant elements in the infrastructure of cyberspace. It is vital to understand their "geography" because they are vulnerable, and difficult and expensive to repair or replace. The complex patterns of their orbital position and surface coverage mean that it is difficult to plot the geography using a single map. One method to chart position and coverage is to use an animated sequence. This approach was taken by Robert Thurman and Patrick Warfolk while at the Geometry Center at the University of Minnesota using their SaVi (satellite visualization) software. Their animations show the changing positions of satellite constellations encircling the globe. The plate bottom-right is a single frame from an animation of the orbital paths of the original design for the Teledesic satellite constellation. The animation shows the "birds" literally marching across the sky in orderly precision. Teledesic is a multi-billion-dollar low-Earth-orbit constellation of several hundred satellites, circling at an altitude of 435 miles and designed to provide broadband data transmission for networks, including the Internet. The individual satellites of the Teledesic constellation are represented by small green dots and their orbital paths are shown by the red tracks. This is overlaid on a simple Earth globe showing country boundaries. The SaVi software can also calculate the footprint of the satellites on the Earth's surface and simulate the view of the satellites overhead from a specified point on the ground.

2.5: International submarine cable infrastructure

chief cartographer: unknown (Alcatel Submarine Systems).
aim: to show the detailed geography of undersea cables for a technical audience.
form: arc–node representation on a base map of coastlines.
technique: large paper map.
date: 1999.
further information: <http://www.alcatel.com/submarine/>

2.6: Frame from a SaVi animation of the Teledesic satellite constellation

chief cartographers: Robert Thurman and Patrick Warfolk (SaVi project, Geometry Center, University of Minnesota).
aim: to show the changing pattern of satellite constellations around the Earth.
form: 3-D globe overlaid with the orbital paths of satellite constellations.
technique: animations produced using the SaVi custom software application.
date: 1997.
further information: SaVi homepage at <http://www.geom.umn.edu/~worfolk/SaVi/>; Teledesic homepage at <http://www.teledesic.com/>
further reading: Lloyd's Satellite Constellations, by Lloyd Wood, <http://www.ee.surrey.ac.uk/Personal/L.Wood/constellations/>

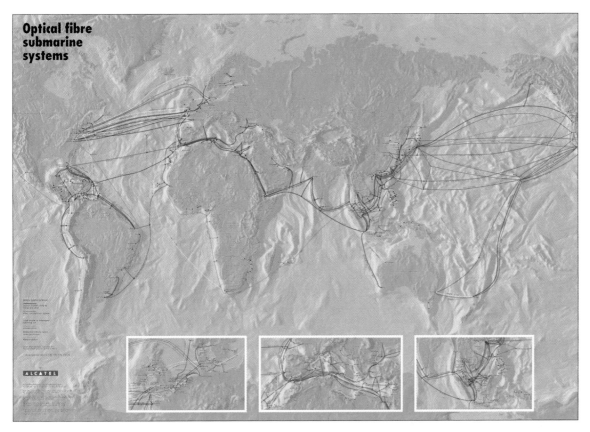

Optical fibre submarine systems

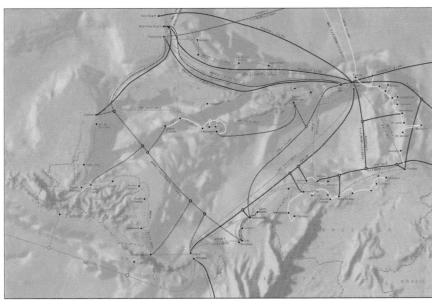

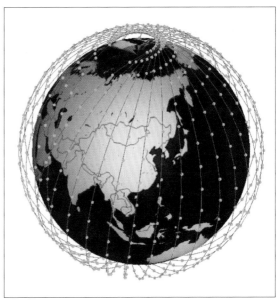

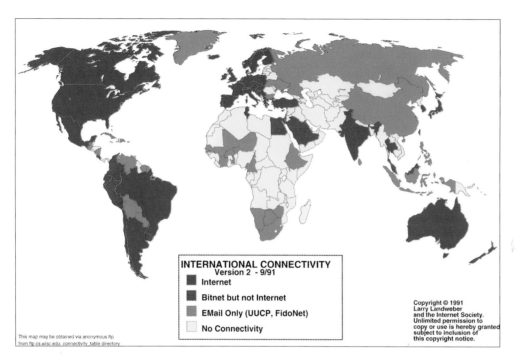

INTERNATIONAL CONNECTIVITY
Version 2 - 9/91

- Internet
- Bitnet but not Internet
- EMail Only (UUCP, FidoNet)
- No Connectivity

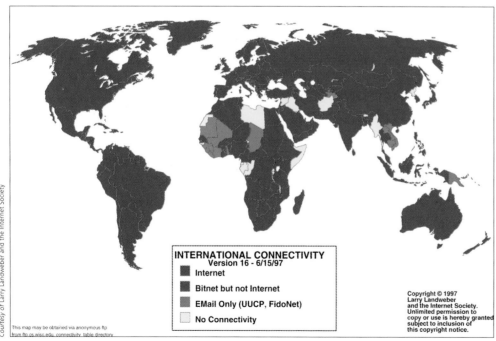

INTERNATIONAL CONNECTIVITY
Version 16 - 6/15/97

- Internet
- Bitnet but not Internet
- EMail Only (UUCP, FidoNet)
- No Connectivity

Infrastructure census maps

The maps presented so far aim to provide information about network topology. In this section we consider three attempts to provide a broader picture of the geographical location of Internet infrastructure. So-called Internet census maps try to provide an overview of the amount of Internet infrastructure (such as the capacity of international links or computers per capita) within countries through the presentation of aggregate-level statistics. The first two examples we present here are both at the global scale, but their techniques have also been used at much smaller scales to demonstrate the unevenness in access to Internet services within cities, states and countries. The third example uses a more disaggregated technique, showing domain name locations at these finer scales.

During the 1990s the Internet spread across the globe, so that by 1997 most nations had some form of connection. This diffusion of the Internet was tracked by Larry Landweber and charted in a series of maps. In total, he produced 12 maps over a period of six years, providing a useful and fascinating visual census of the spread of international network connectivity. The top map displays national-level diffusion of the Internet as at September 1991. The bottom map is the last in the Landweber series and was created in the summer of 1997. In each of the maps a fourfold classification is used to represent network connectivity. No connectivity is represented by yellow; the two intermediate connectivity levels – to email only and to BITNET – are represented by green and red respectively; and the highest category of permanent Internet connectivity, with a full range of interactive services, is represented in blue. Data on the changing state of network connectivity in different countries were gathered by Landweber from a network of human contacts across the world.

These two maps reveal a partial picture of global Internet diffusion through the 1990s. The first, from 1991, shows that a large number of countries, particularly in the Americas and in Northern Europe, had full Internet connectivity. However, an equally large measure of the world's nations are shaded yellow, indicating that they had no Internet connectivity. In fact this category included well over half the nations of the world, though these were clearly concentrated in the less-developed regions of Africa and central Asia. By 1997, the majority of the nations of the world were shaded blue: the Internet, as measured by Landweber's survey, was so widespread that the exceptions really stand out. (It was at this point that tracking diffusion at this scale became redundant and, hence, this is the last map in the series.) The yellow-shaded exceptions were nations suffering from extreme poverty, war and civil conflicts (such as Afghanistan and Somalia) or from geopolitical isolation (e.g. Libya, North Korea, Burma, Iran and Iraq).

Landweber's maps represent data by shading countries according to a classification scheme. This is a very common cartographic-design approach known as choropleth mapping, which is widely used to map statistical data. It is easy to assume that these maps provide a clear and straightforward geographic presentation of the data. However, this is not always the case, and one must interpret them carefully to avoid drawing naive and unsound conclusions about the patterns in the data. In the case of the International Connectivity series of maps, an uncritical reading of them could easily provide a distorted view of the global spread of the Internet. As discussed in full in chapter 1, whilst most of the world is connected to the Internet, this level of connectivity is not equally distributed in scope or cost.

2.7: International Connectivity series of maps
chief cartographer: Larry Landweber (Computer Science Department, University of Wisconsin-Madison).
aim: to chart the global spread of network connectivity at the level of nation-states.
form: choropleth maps where countries are shaded according to which one of four categories of network connectivity they fit.
technique: digital maps available as Postscript and bitmap images.
dates: the first map was created in September 1991 and the last one in June 1997.
further information: all the maps and supporting data tables are available from <ftp://ftp.cs.wisc.edu/connectivity_table>

The longest-serving cartographer of the Internet is undoubtedly John S. Quarterman. Through his research consultancy, Matrix.Net, based in Austin, Texas, he has been actively measuring, analyzing and mapping the geography of the Net for the past decade or more. His company makes many different maps of the Internet, but perhaps the most significant is the world map showing the whole Internet in one snapshot (also see the Internet Weather maps, page 66). The examples here show the state of the Internet in January 1997 and January 2000 respectively. They provide an overview of the global Internet by mapping the aggregated volume of networked computers, known as "hosts", using the common cartographic design of graduated circles; the larger the circle, the more hosts are located in that geographic element, where the circles are plotted onto the familiar geographic framework of continents and countries. The map opposite from January 1997 displays three other global computer networks (BITEARN, UUCP, and FidoNet) in addition to the Internet. These networks have declined dramatically in use as the Internet has come to dominate; hence, they have been dropped from later Matrix.Net world maps such as that shown below.

The maps provide a more detailed overview than the Landweber maps, revealing the extent to which Internet infrastructure is distributed within countries, and the large concentrations in North America, Europe and East Asia. The large number of purple circles in North America, representing a million or more Internet hosts, demonstrates the extent to which this area of the world still dominates Internet usage – although Europe and East Asia are clearly catching up, as evidenced by their clusters of blue circles. Unlike Landweber's maps, then, a north–south divide is clearly evident, with developing countries in Africa and South America having comparatively fewer hosts. Further interpretation, beyond a global overview, is difficult because the scale of the map has led to significant overplotting, making it difficult to identify the geographic locale of each symbol.

2.8: Matrix.Net Internet world maps

chief cartographer: John S. Quarterman and his colleagues (Matrix.Net, Inc., Austin, Texas).
aim: to chart the geographic extent of the Internet as a function of the volume of networked computers at the city level. The first map charted four different networks, while the later one shows only Internet-connected computers.
form: world map with networked computers represented by a graduated circular symbol.
technique: digital maps as bitmaps and Postscript-generated using custom software and mapping application.
dates: January 1997, January 2000.
further information: Matrix.Net homepage at <http://www.matrix.Net>

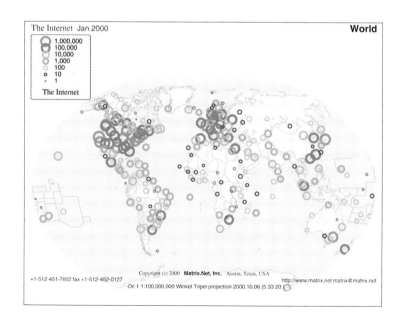

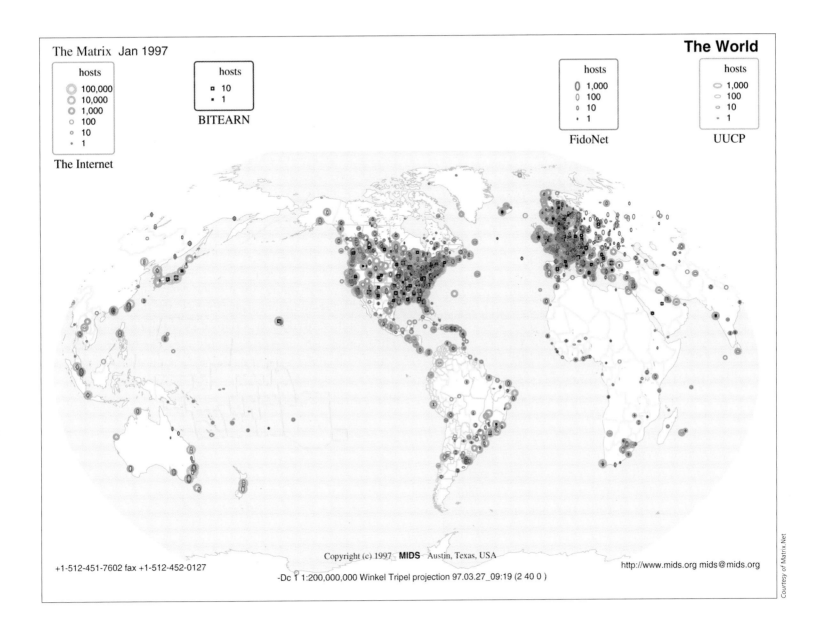

The Matrix Jan 1997

The World

hosts
◎ 100,000
◎ 10,000
◎ 1,000
○ 100
○ 10
· 1

The Internet

hosts
▫ 10
▪ 1

BITEARN

hosts
0 1,000
0 100
○ 10
· 1

FidoNet

hosts
◯ 1,000
◯ 100
○ 10
· 1

UUCP

+1-512-451-7602 fax +1-512-452-0127

Copyright (c) 1997 **MIDS** Austin, Texas, USA

-Dc 1 1:200,000,000 Winkel Tripel projection 97.03.27_09:19 (2 40 0)

http://www.mids.org mids@mids.org

Courtesy of Matrix.Net

Domain name maps

Computers identify other computers on the Internet using a globally unique number called an IP (Internet Protocol) address. IP addresses are 32-bit numbers consisting of four numbers ranging from 0 to 255, separated by a period (e.g., 144.82.100.130). The particular IP number in the foregoing example identifies a single Unix workstation at University College London, providing it with a globally unique location on the Internet. Numeric IP addresses were originally the only form of virtual address. However, they proved difficult for people to remember and so an alternative system of addressing, called "domain names", was developed. Domain names are short textual names (e.g., www.ucl.ac.uk) that are structured in a hierarchical system, which can be thought of as a tree that grows upwards from the general to the specific. Domain names have become an increasingly valuable commodity as corporations seek memorable addresses for their websites.

There have been a number of attempts to map and analyze the geographic distribution of ownership of domain names at different scales. This task, however, has had two main problems. First, names are allocated by many different agencies; and, second, it is difficult to pin down the exact size of global domain names because of their delegated structure and rapid growth. (As of November 2000 there were over 33 million domain names registered, of which 19.9 million were .com). To date, most maps have concentrated on plotting the ownership of .com domains in the United States. The geographic location of the owner of these domains can be determined from the registration database, which has a billing postal address containing zip codes that can quite easily be mapped to street-level locations.

Matthew Zook, in the Department of City and Regional Planning at University of California, Berkeley, has produced a series of domain name maps, from a regional scale down to the scale of street in a number of selected cities. This is part of his doctoral research into the geography of the Internet industry. Zook gathered his own data on the ownership location of .com, .net, .org and .edu domains from InterNIC using a computerized survey that ran for five weeks in the summer of 1998. Three of Zook's maps for the San Francisco region are displayed opposite.

Zook uses simple dot and proportional symbol maps, with background road and town data to add context. These illustrate well the potential of this type of Internet census-mapping. Each symbol represents the number of domain names at that particular zip code. The top map is at the regional level, showing the Bay area with dense clusters evident in San Francisco itself, down Silicon Valley and in the Oakland/Berkeley area. The bottom-left map displays the domain names in San Francisco itself, with the densest concentrations in the financial district and "South of Market" area (famed as "Multimedia Gulch"). The bottom-right map displays only a few city blocks around South Park, at the heart of the multimedia district. A high degree of geographic clustering of domains is clearly visible across the three spatial scales.

2.9: Domain name maps for the San Francisco region

chief cartographer: Matthew Zook (Department of City and Regional Planning, University of California at Berkeley).
aim: to map the distribution of domain name ownership.
form: graduated dot map on base data, where the size of the symbol is proportional to the number of domains at a location.
technique: digital maps produced using geographical information systems, based on domain name registration data.
date: 1998.
further information: Internet Geography Project at
<http://socrates.berkeley.edu/~zook/domain_names/>

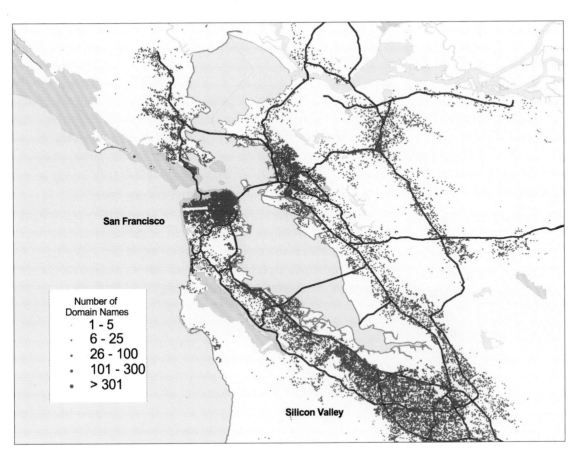

San Francisco

Number of
Domain Names
· 1 - 5
· 6 - 25
· 26 - 100
· 101 - 300
● > 301

Silicon Valley

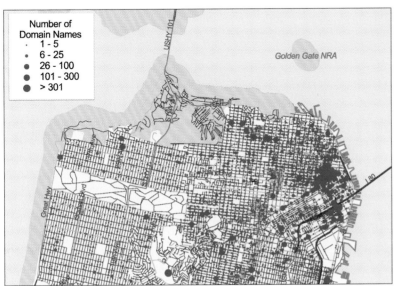

Number of
Domain Names
· 1 - 5
· 6 - 25
● 26 - 100
● 101 - 300
● > 301

Golden Gate NRA

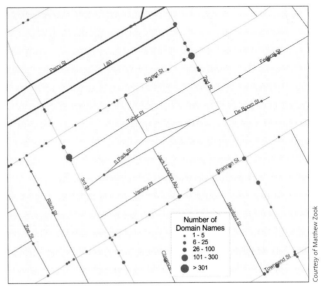

Number of
Domain Names
· 1 - 5
· 6 - 25
● 26 - 100
● 101 - 300
● > 301

Courtesy of Matthew Zook

Marketing maps of Internet service providers

A large number of maps of the Internet have been produced by commercial network operators for the purposes of marketing. Here, maps serve an overt commercial purpose, namely to attract potential customers in what is a highly competitive and lucrative global business. The maps created seek to communicate to potential customers the benefits of using the network operators' services by highlighting two key aspects of their infrastructure: its geographic extent and its capacity.

UUNET is one of the world's leading Internet network providers, being part of the massive MCI WorldCom corporation. UUNET's network maps are designed to promote the impressive capacity and connectivity that the company offers in different regions of the world. They are created by Henry Ritson, UUNET's global marketing manager, as an integral part of UUNET's sales and marketing strategy. The maps are produced at four different scales – global, continental, national and city – and are updated on a quarterly basis using information passed on from the engineering department. This information is turned into high-quality, hand-crafted maps, a time-consuming cartographic exercise taking several days' work for each map.

The cartographic design of Ritson's marketing maps opposite uses conventional arc–node topology to represent the network on a geographic base. In the global and continental maps, network links between cities are shown by lines where their color and thickness encode capacity measured in millions of bits transferred per second (Mbps). The thin purple lines are 45 Mbps (known as T3s) and the range goes up to thick black lines that are the 12-lane superhighways of the Net carrying a massive 2,488 Mbps. The use of the geographic base map provides an important and familiar template for users, illustrating the reach of the network. As such, the maps provide an abstracted picture of UUNET's network, using a process of generalization to produce relatively neat and ordered maps suitable for marketing purposes. However, they only capture the state of UUNET's network at a single point in time. In the frenetic world of Internet infrastructure provision, UUNET's network is reportedly growing at 1,000 percent a year, and links and nodes are added and upgraded all the time. The engineers who design, build and maintain the network can obviously track the infrastructure, but the maps and schematics they use for this task are full of highly technical and confidential information and are therefore not publicly available.

It is clear from these marketing maps that UUNET has developed an extensive, high-capacity network forming a dense mesh of links connecting many parts of the world. Indeed, the density of network capacity in Silicon Valley and along the Washington–Boston corridor means that these areas on the map have limited cartographic legibility.

2.10: UUNET's backbone marketing maps at four different scales
chief cartographer: Henry Ritson (UUNET), UK map artwork by Mark Watts.
aim: to provide marketing maps to promote the company's Internet network to potential customers by demonstrating the geographic extent and capacity of the infrastructure.
form: high-quality arc–node representations mapped onto a geographic framework, with symbol size and color denoting different capacities of links and hubs. Different scale maps reveal more detail about network topology and capacity.
technique: hand-crafted maps, updated quarterly from detailed information concerning network engineering.
date: global – June 2000; North America – June 2000; UK/London – October 2000.
further information: UUNET Network homepage at <http://www.uu.net/network> UUNET UK at <http://www.uunet.co.uk/network>.

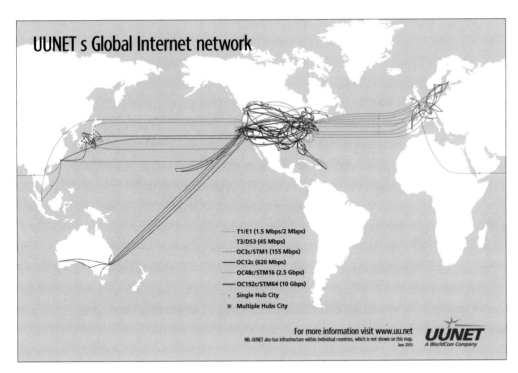

UUNET s Global Internet network

T1/E1 (1.5 Mbps/2 Mbps)
T3/DS3 (45 Mbps)
OC3c/STM1 (155 Mbps)
OC12c (620 Mbps)
OC48c/STM16 (2.5 Gbps)
OC192c/STM64 (10 Gbps)
- Single Hub City
- Multiple Hubs City

For more information visit www.uu.net
NB: UUNET also has infrastructure within individual countries, which is not shown on this map.
June 2000

UUNET
A WorldCom Company

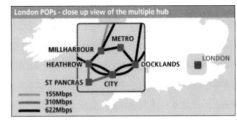

London POPs - close up view of the multiple hub

MILLHARBOUR METRO
HEATHROW DOCKLANDS LONDON
ST PANCRAS CITY

155Mbps
310Mbps
622Mbps

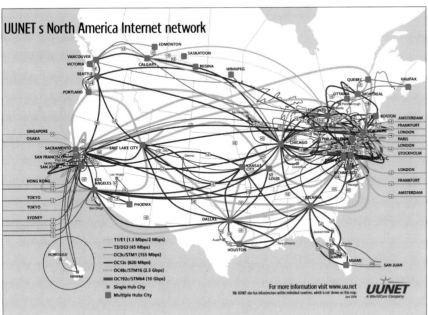

UUNET s North America Internet network

T1/E1 (1.5 Mbps/2 Mbps)
T3/DS3 (45 Mbps)
OC3c/STM1 (155 Mbps)
OC12c (620 Mbps)
OC48c/STM16 (2.5 Gbps)
OC192c/STM64 (10 Gbps)
- Single Hub City
- Multiple Hubs City

For more information visit www.uu.net
NB: UUNET also has infrastructure within individual countries, which is not shown on this map.
June 2000

UUNET
A WorldCom Company

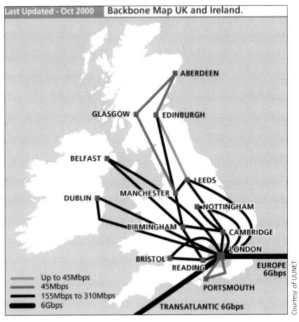

Last Updated - Oct 2000   Backbone Map UK and Ireland.

Up to 45Mbps
45Mbps
155Mbps to 310Mbps
6Gbps

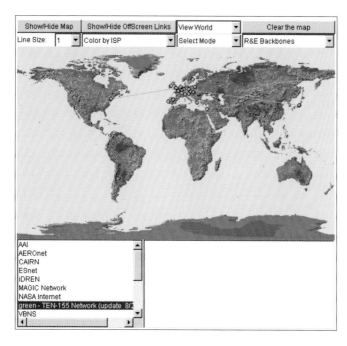

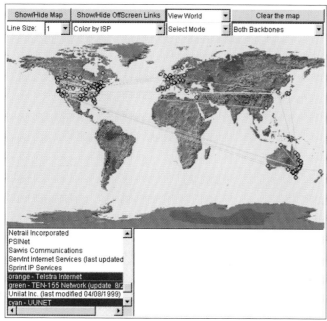

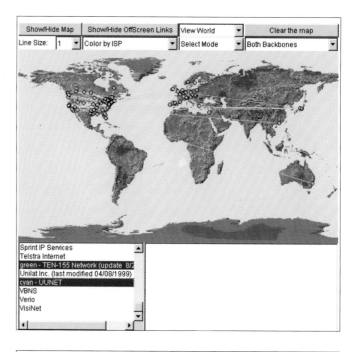

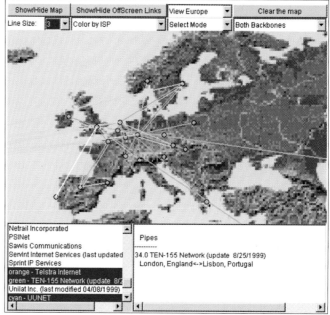

Interactive mapping of networks

The maps we have discussed so far convey information in a fixed form that cannot be altered or queried by the map reader. In contrast, some researchers are exploring more sophisticated representations that allow the map reader to interact with the map and to explore the visualization further. These maps shift some of the power of representation from map author to map viewer, allowing the user to explore the dataset (although this is within the parameters set by the map's creator).

Interactive maps can take a number of forms. Some allow the map reader to change viewing position, others to modify the visual presentation by changing the data classification or symbology, or through the subsetting of an area of interest by zooming or database selection, and others still to enquire interactively about individual data objects. In some cases, these features are combined. All of these approaches are pushing the conventional boundaries of map representation into the realm of cartographic visualization. Moreover, they often utilize the power of the Web to deliver data and interactive representations onto a user's desktop, where users are then free to explore them at will.

Mapnet is an interactive mapping tool that allows Web users to map and examine multiple Internet-backbone infrastructures from more than 30 different commercial and education-and-research networks. Mapnet was developed by Brad Huffaker at CAIDA and is programed in Java so that it can be run using a standard Web browser. The plates opposite display a series of screenshots of Mapnet in action, mapping the network infrastructures of the Ten-155 network in Europe (shown by green links), then adding UUNET (cyan links) and then Telstra's network centered in Australia (orange links). Mapnet visualizes networks as arc–node topologies on a flat, terrain-style, geographic base map. The interface comprises four major elements: in the center is the large map window; above this is a series of control buttons and menus that provide a number of useful interactive functions; and at the bottom are two text boxes, the one on the left providing a selection list of networks to map and the one on the right displaying the results of interactive queries.

The Mapnet application offers a significant degree of control over data representation and enquiry. One can set the line thickness and color-code the arcs by company or by bandwidth. It is also possible to zoom from the global view to regional levels to see detail more clearly, as the bottom-right screenshot shows by zooming into Europe. Moreover, using the enquiry mode it is possible to click on a link of interest and receive details about that link in terms of the end nodes, capacity, and company that owns it. In the example, the link between London and Lisbon is highlighted.

Mapnet thus provides the user with the flexibility to explore the network infrastructures of their choice, mapped onto a common, familiar framework of real-world geography. This is important because it allows the comparison of networks of different companies. Usually it is difficult to compare networks because information is generally only available in marketing maps, which are presented in all manner of different formats and cartographic styles. This said, Mapnet is reliant on commercial networks to provide up-to-date information – something many are reluctant to do.

2.11: Mapnet, an interactive Internet-backbone mapping tool

chief cartographer: Brad Huffaker (Cooperative Association for Internet Data Analysis).
aim: to provide an interactive tool that allows users to explore geographic extent and bandwidth of multiple Internet networks.
form: geographic world map with networks drawn as arc–nodes that are color-coded by owner or bandwidth.
technique: interactive Java program that allows users to choose which network to map, how to color-code, and how to zoom in to see regions in more detail.
dates: Mapnet went online in 1997 (screenshots opposite taken in May 2000).
further information: CAIDA, homepage at <http://www.caida.org>
Mapnet available at <http://www.caida.org/tools/visualization/mapnet/>

These plates show three different views of a three-dimensional, interactive map of the backbone network of CESNET, the education-and-research Internet network of the Czech Republic, as at May 1997. The network links between nodal towns and cities are represented by the horizontal tubes, with the width of the tube indicating the capacity of that link. The central focus of the network is the capital Praha (Prague), which is clearly identifiable as the center of the strong star-shaped configuration. This map is a 3-D model, constructed using Virtual Reality Modelling Language (VRML). When displayed using a suitable VRML viewer, the data are easily manipulated and viewed from any angle. The user can therefore freely rotate, zoom, pan and flip the 3-D backbone map, as can be seen in the three screenshots.

The top plate opposite shows the image viewed from its default position, "face on", and it looks like a conventional flat 2-D map. In the bottom image the map has been tilted on its side to reveal the 3-D arc–node structure. In the final image below the viewing position is as if the model had been flown into, presenting the effect of zooming into part of the network (in this case the Prague area). In some senses, then, the map reader can "get inside" the map. Constructing the map data as a 3-D model means that it can be examined from an almost unlimited number of positions and angles, giving more power to the viewer as compared with fixed or even animated maps.

2.12: 3-D map of the CESNET Internet backbone

chief cartographer: unknown (CESNET).
aim: to map the topology of the CESNET backbone in the Czech Republic.
form: a 3-D map with network links as pipes floating above an outline map of the Czech Republic. City nodes are shown by small columns.
technique: geographic arc–node representation of the network using VRML to provide a 3-D model that can be viewed from any angle.
date: May 1997.
further information: CESNET homepage at <http://www.cesnet.cz>

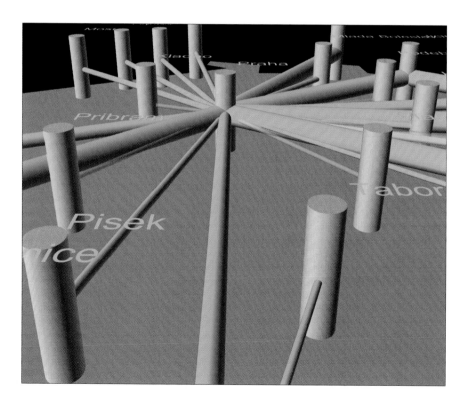

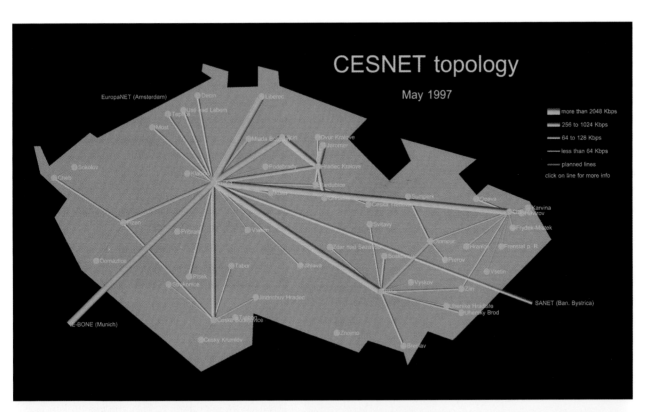

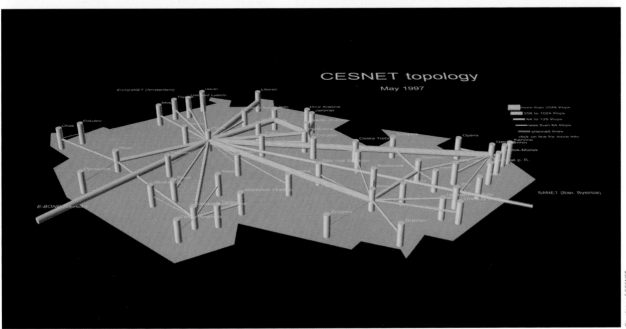

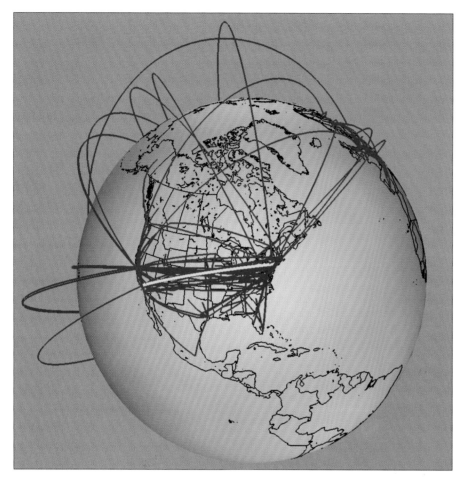

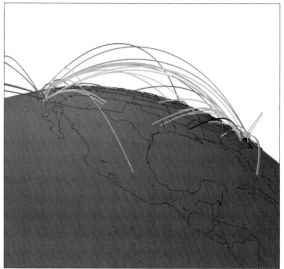

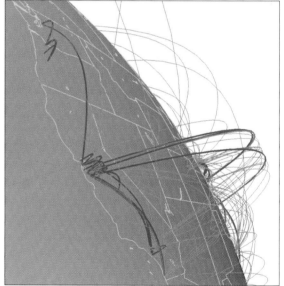

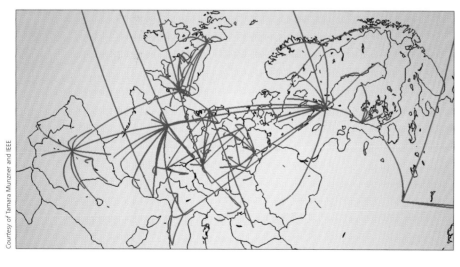

The globe is one of the most powerful visual metaphors onto which to map the geographical dimensions of large networks such as shipping lines, trade flows, airline routes and communications. A team of researchers, Tamara Munzner, K. Claffy, Eric Hoffman and Bill Fenner, have utilized this metaphor to produce visually striking interactive maps of a significant part of the Internet – the MBone, which comprises a special set of routes, known as "tunnels" in technical Internet-speak, that run on top of ordinary Internet networks and are used to deliver multicast data. Multicasting is an Internet protocol designed for efficiently delivering a single copy of a chunk of data to many different people. It is especially useful for distributing real-time audio and video communications, such as of live concerts or space shuttle launches, to a large audience without sending individual copies to everyone. The tunnels themselves are created between special routing computers, forming a dedicated multicast backbone network, known as MBone for short.

From mid-1996 the cartographers have represented the MBone using an arc–node representation projected into a 3-D space. The globe around which these arcs are placed is marked with country boundaries, along with US state and Canadian provincial borders, to provide a geographic context. Curving lines between MBone router locations have been employed to represent the 700-odd MBone tunnels, with color and thickness used to show characteristics of the tunnels while the height of the arcs above the surface of the globe was simply a function of distance. The longest tunnels were drawn as the highest arcs, as can be seen opposite. This makes sense in the context of understanding the MBone topology, because long links that span large parts of the Internet are the most important and so merit greater visual prominence. It should be noted that these MBone visualizations only show the structure of this network and not any data on traffic actually flowing over the links.

One of the aims in the researchers' MBone mapping endeavors was to make the results as widely available as possible and, also, to make the maps as interactive as possible, so that people could explore the topology for themselves rather than being presented with predetermined static images. To achieve these aims the researchers produced their 3-D maps as interactive VRML models. The maps opposite present a series of impressive screenshots of their 3-D MBone maps from a range of viewpoints. The top-right map is a view of the whole globe, as if seen from several hundred miles out in space, with a dense mesh of arcs criss-crossing the United States from coast to coast, along with higher and longer transcontinental tunnels curving around the globe. One particular MBone link has been selected and is highlighted as a yellow arc. The map bottom-right shows the European MBone structure.

The remaining two smaller figures show horizon views of different aspects of the United States. The map top-left is a view of Mexico and the southern states from a vantage point above Central America. In this example, the MBone links have been color-coded according to the company that owns the network links. The map bottom-right is a view taken from a low angle, close to the surface of the globe off the coast of California. Looking across North America from this angle, the density of looping arcs of varying heights is clearly apparent. Lateral links along the west coast, as well as longer tunnels from the east coast, can be discerned. Certain tunnels are highlighted using a different color and thickness of lines, indicating that they are of a different specification.

2.13: 3-D MBone globes

chief cartographers: Tamara Munzner (Computer Graphics Laboratory, Stanford University), K. Claffy (CAIDA), Eric Hoffman and Bill Fenner.
aim: to visualize the topological structure of the part of the Internet called the "MBone" on a geographic base.
form: 3-D arcs on a globe, with country boundaries shown to provide context.
technique: custom software to create 3-D visualization distributed in VRML.
date: June 1996.
further information: Planet Multicast homepage at <http://oceana.nlanr.net/PlanetMulticast/>
further reading: "Visualizing the global topology of the MBone", by Tamara Munzner, K. Claffy, Eric Hoffman and Bill Fenner, proceedings of the 1996 IEEE Symposium on Information Visualization, 28–29 October 1996, San Francisco, pages 85–92. <http://www-graphics.stanford.edu/papers/mbone/>

Visualizing network topologies in abstract space

In this section, we turn our attention away from attempts to map the Internet's infrastructure onto geographic space to examining more abstract spatializations. These are essentially topological maps of the infrastructure that have been visualized into forms that do not correspond to geographic space. The spatializations most often use graph-type structures as their basic visual metaphor, with a number of different layout styles. Here, the absolute, geographical location of the infrastructure is not important, and the spatializations are designed to reveal other kinds of information using a system of relative location. For example, the spatializations might reveal connectivity and routing that may well be indiscernible on a cluttered geographic representation.

As with other maps of infrastructure, these spatializations take on a number of forms and are used by different groups of people; some spatializations are technical blueprints used by those who manage the networks; others are aimed at the users of the networks.

The spatialization shown opposite was created by Elan Amir whilst a graduate student in the Computer Science Division at the University of California at Berkeley. The spatialization shows some 1,377 routers on the MBone in August 1996 as an organic-looking graph with dense clusters and many scattered outlying limbs. It looks almost astronomical, with the MBone nodes and connections floating free in an abstract space, like stars in distant galactic clusters. This graph can be thought of as a non-geographic version of the MBone maps of Tamara Munzner and colleagues (see previous page).

The map was created using a network drawing tool developed by Amir, called Carta. MBone routers are represented by boxes labeled with an ID number. Whilst having limited practical use, the spatialization provides an interesting holistic view of the MBone topology from a single view, providing a sense of the shape and interweaving structure of that part of the Internet infrastructure.

2.14: Large graph of the MBone network topology

chief cartographer: Elan Amir (while at the Computer Science Division, University of California, Berkeley).
aim: to visualize the topological structure of the MBone.
form: a complex, organic-looking graph, laid out in an abstract space.
technique: 2-D postscript graph created using Amir's Carta software.
date: August 1996.
further information: <http://www.cs.berkeley.edu/~elan/mbone.html>

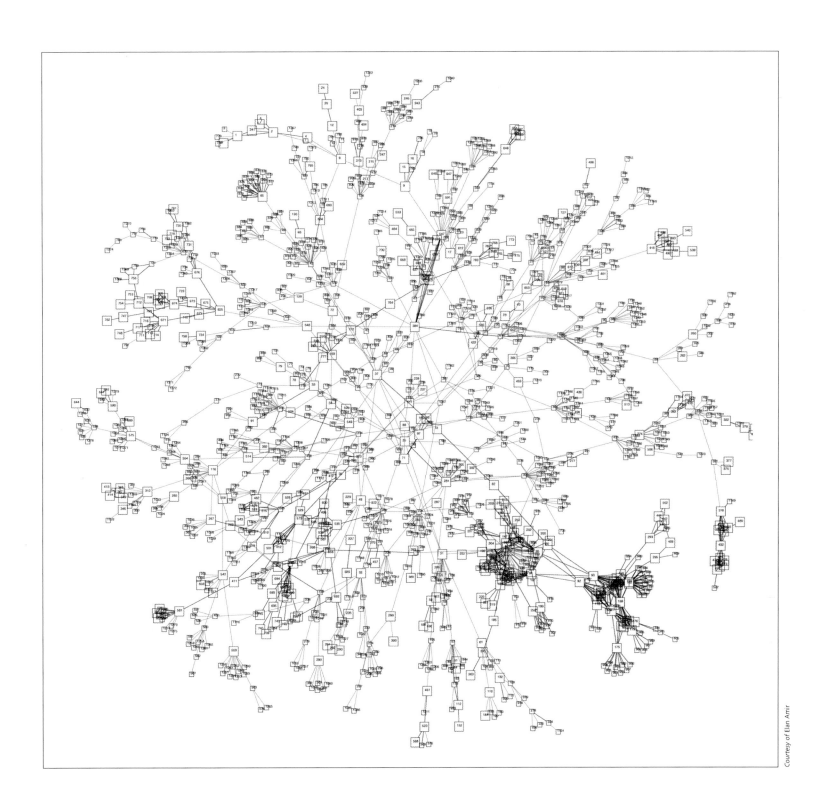

Mapping infrastructure and traffic **39**

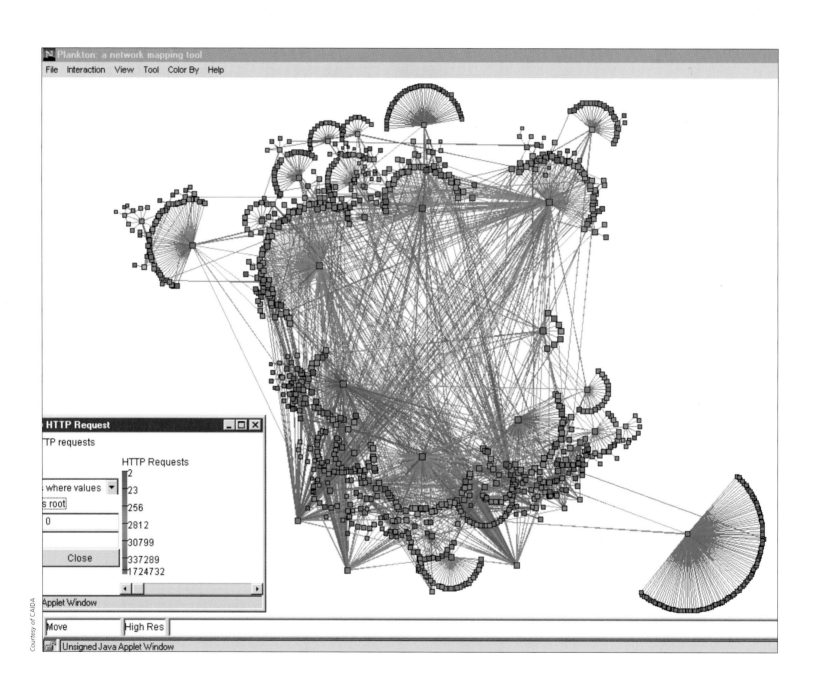

Plankton: a network mapping tool

File   Interaction   View   Tool   Color By   Help

HTTP Request

TP requests

HTTP Requests
2
where values   ▼   23
s root   256
0   2812
                          30799
Close   337289
                          1724732

Applet Window

Move   High Res

Unsigned Java Applet Window

Courtesy of CAIDA

Plankton is an interactive network mapping tool developed to create topological maps of a particular and important element of Internet infrastructure, namely international Web caches. These caches help make the Web work more efficiently and improve the speed for end users by storing commonly sent data. The screenshots here show Plankton in action, with each small square representing a cache node and the lines showing how they are connected into hierarchies. The inset image below displays a small part of this complex hierarchy in finer detail. The nodes and lines are colored according to traffic demand, with darkest red being the busiest. Data on cache topology are gathered daily, and users are able to select which day they want to examine.

Plankton provides a number of interactive functions, including graph layout (geographic or abstract), symbology (node size, link thickness, or color-coding by traffic or domain), rotating, panning, zooming and the creation of time-series animation. In addition, individual nodes can be interrogated.

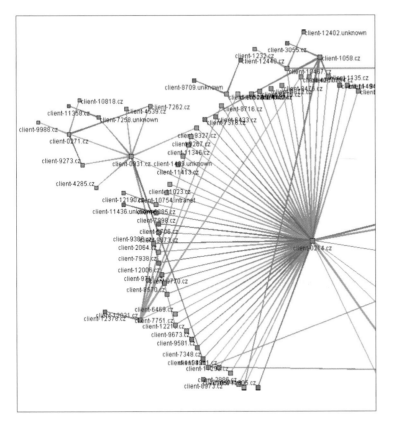

2.15: Plankton, a visualization of the topology of global Web caches

chief cartographers: Brad Huffaker, Jaeyeon Jung and K. Claffy (Cooperative Association for Internet Data Analysis – CAIDA).
aim: to create a spatialization of global Web caches.
form: hierarchical graphs drawn in semicircular configurations, color-coded by traffic.
technique: interactive Java application.
date: application developed in 1997, screenshot captured with June 2000 cache data.
further information:
Plankton homepage at <http://www.caida.org/tools/visualization/plankton/>
further reading: Visualization of the Growth and Topology of the NLANR Caching Hierarchy, by Bradley Huffaker, Jaeyeon Jung, Duane Wessels, and K. Claffy, CAIDA Report, March 1998. <http://www.caida.org/tools/visualization/plankton/Paper/plankton.html>

In terms of attempts to map the whole Internet, a rival form to the Internet census maps has emerged with the work of Bill Cheswick and Hal Burch. Their Internet Mapping Project spatializes the topology of thousands of interconnected Internet networks to provide an overview of the core of the Internet in a single snapshot. They map the Internet in an abstract space: as Cheswick argues, "We don't try to lay out the Internet according to geography … The Internet is its own space, independent of geography."

Data are gathered by using the Internet to measure itself on a daily basis, surveying the routes to a large number of end points (usually Web servers) from the research project's base in New Jersey. The resulting spatializations reveal how the hundreds of networks and many thousands of nodes connect to form the core of the Internet. The striking example shown opposite is a spatialization of data gathered on 11 December 2000, representing nearly 100,000 nodes.

The spatialization takes several hours to generate on a typical PC. The layout algorithm nevertheless uses simple rules, with forces of attraction and repulsion jostling the nodes into a stable, legible configuration. The end result is a static image, but there are many permutations in the algorithm to generate different layouts and color-codings of the links according to different criteria (such as network ownership or country). In the example shown, links have been color-coded according to the Internet service provider (ISP), seeking to highlight who "owns" the largest sections of Internet topology.

This project is ongoing and the data are archived and available to other researchers to utilize. Over time, it is hoped that the data will be useful for monitoring growth and changes in the structure of the Internet.

2.16: Internet connectivity graph

chief cartographers: Hal Burch and Bill Cheswick (Lumeta).
aim: to visualize the core of the Internet in a single graph.
form: a striking graph that has been variously described as a peacock's wing, a lung or a coral reef. It has an organic look of fractal complexity. Color-coding identifies networks of major operators.
technique: Internet structure measured daily and maps rendered using custom graph-drawing software which takes many hours to lay out the final image.
date: December 2000.
further information: <http://www.lumeta.com>
Internet Mapping Project <http://www.cs.bell-labs.com/~ches/map/index.html>
Peacock Maps <http://www.peacockmaps.com>

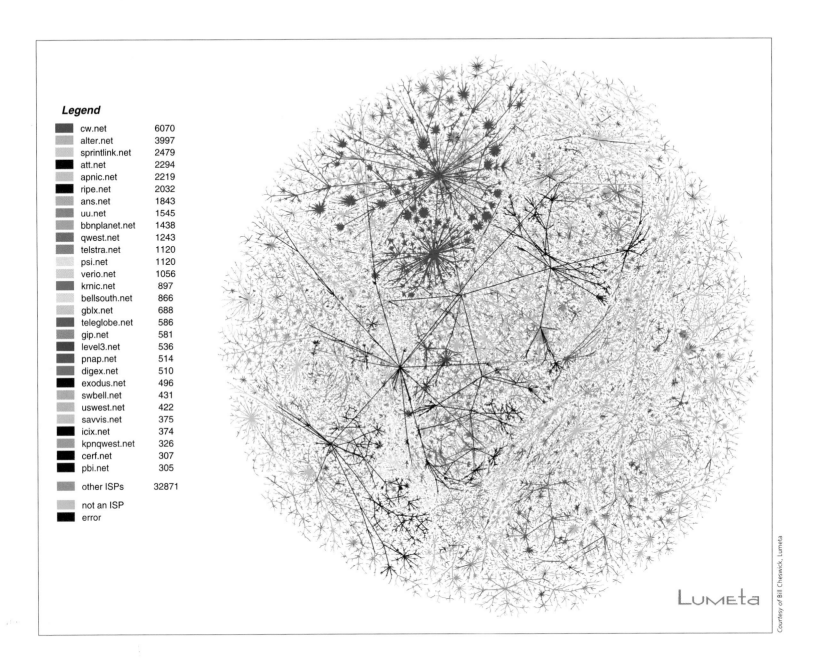

	cw.net	6070
	alter.net	3997
	sprintlink.net	2479
	att.net	2294
	apnic.net	2219
	ripe.net	2032
	ans.net	1843
	uu.net	1545
	bbnplanet.net	1438
	qwest.net	1243
	telstra.net	1120
	psi.net	1120
	verio.net	1056
	krnic.net	897
	bellsouth.net	866
	gblx.net	688
	teleglobe.net	586
	gip.net	581
	level3.net	536
	pnap.net	514
	digex.net	510
	exodus.net	496
	swbell.net	431
	uswest.net	422
	savvis.net	375
	icix.net	374
	kpnqwest.net	326
	cerf.net	307
	pbi.net	305
	other ISPs	32871
	not an ISP	
	error	

Courtesy of Bill Cheswick, Lumeta

Moving downscale from spatializations that try to show the "whole" of the global Internet, we now present an example that focusses on a single territory. The "Cyber Map" of Hong Kong provides a comprehensive visual census of all Internet backbone connections between ISPs in Hong Kong and their international links. It is a hard-copy paper poster-sized spatialization and is provided as a free supplement to IDG's computer magazine in Hong Kong. The data on ISP interconnections is laboriously gathered and verified by hand unlike, say, Cheswick and Burch's Internet Mapping Project just examined, which uses automatic measurement probes.

In style, the spatialization is an abstract topological representation reminiscent of old-fashioned astrological star charts. It uses a simple and effective symbology of circular nodes representing hubs of different ISPs and lines for the interconnecting links. Strong color-coding of the lines and circles is used to represent details about the different types of links and the nature of the ISP. Labels also show the name and/or abbreviation of the ISP and the bandwidth of the link. International hubs are arranged in boxes at the top and bottom of the spatialization; the heart of the diagram is the Hong Kong Internet Exchange (HKIX – http://www.hkix.net/), which is the key hub and provides a "neutral" meeting point where a large number of competing ISPs can interconnect to share traffic. The HKIX is shown on the spatialization by a double-ringed white circle, which appears to be the brightest "star" at the center of Hong Kong's Net constellation. For Internet users, the spatialization provides a comprehensive and comprehensible survey of Internet topology and capacity.

2.17: "Cyber Map" of Internet interconnections in Hong Kong
chief cartographer: Ernest Luk (IDG Communications, Hong Kong).
aim: to show the network topology and capacity for all interconnections in Hong Kong and to international networks.
form: abstract topological map using simple nodes and arcs.
technique: manual survey and construction.
date: December 1999, updated quarterly.
further information: <http://www.idg.com.hk/cybermap/>

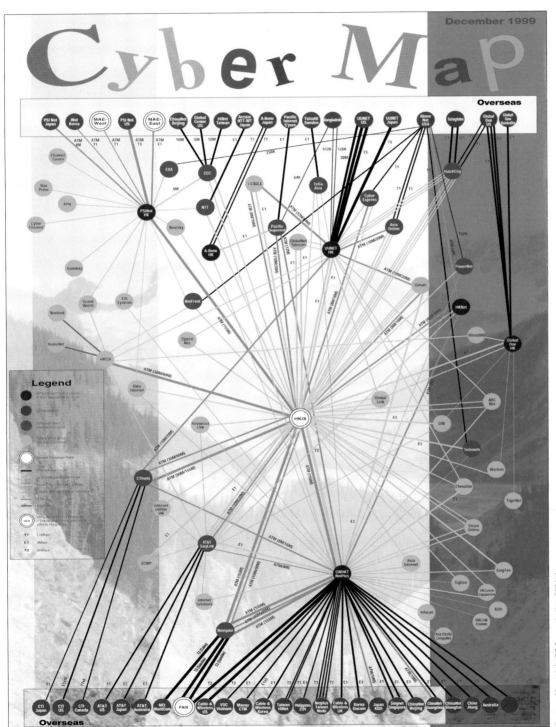

December 1999

Cyber Map

Courtesy of Ernest Luk, IDG Communications (HK) Ltd

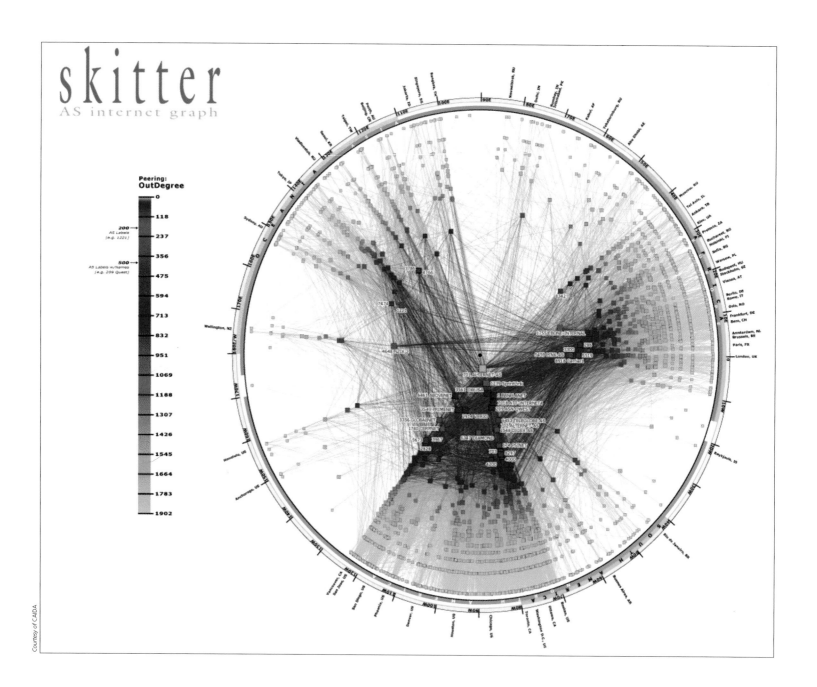

skitter
AS internet graph

Peering:
OutDegree

This graph is an intriguing summarization and visualization of the complex topology of the core of Internet connections. It is based on data gathered by CAIDA's skitter program and Bill Cheswick's Internet connectivity project (see page 42). The aim was to show how much ISPs were interconnected in terms of their peer relationships and how they are dispersed around the world. Importantly, the researchers who created this visualization aggregated and filtered the data first, rather than trying to visualize the whole dataset, as occurs with Lumeta's 2-D Internet graphs or the 3-D hyperbolic graphs discussed next. The 220,000 individual nodes in the dataset, of Internet topology were aggregated by the ISP to which they belonged, based on the technical grouping by autonomous system (AS) number. This yielded some 1,516 AS nodes, which represent the most highly connected ISPs that form the core of the Internet and carry the bulk of the traffic.

These data were then visualized as a polar-projected graph, where each AS node was represented by a small square. The spatial position of the squares encode two vital characteristics: first, the distance away from the center shows the relative strength of peer relationships with other ISPs; second, the angular position around the circumference of the circle shows its approximate geographic position. The links between ISPs are shown by the arcs, which are color-coded "hot" (red) to "cold" (blue) based on the relative strength of connection, so yellow, central, nodes are some of the most well-connected ISPs.

In terms of the geographic component of the visualization, one can think of this as somewhat like a map of the Earth with a projection centered on the North Pole. Around the circumference of the graph, longitudes are marked every ten degrees, and colored strips denote the different continents. Key cities are also labeled. The geographic position of an AS node is determined by the headquarters location of the ISP. Clearly, there is considerable generalization because many large ISPs will have their infrastructure spread across the globe. As such, one might like to think of this as more of a geopolitical location. The graph generally divides the nodes into three distinct segments based on the continents of North America, Europe and Asia/Oceania.

The key aspect of the AS Core Internet graph is its ability to identify the most powerful ISPs and where on the globe they are concentrated. (In this visualization – powerful in the sense that power for ISPs comes through peering – the more peer linkages with other ISPs, the more efficiently you will be able to route your customers' data.) It is clear that the densest concentration of AS nodes, towards the center of the graph, lies along the longitude of about 70° West, which relates to the eastern seaboard of the USA. The headquarters of some of the leading Internet backbone operators, such as UUNET, PSInet, Qwest and CWUSA, are found there.

In many ways this visualization reinforces what is apparent from many of the other maps in this chapter: that the United States is the dominant player in terms of Internet infrastructure. In fact, the top 15 most connected AS nodes are in North America (just one being in Canada). The AS Core Internet graph also reveals that many of the ISPs based in Europe and Asia–Oceania have relatively far fewer connections between them, relying instead on peer links with US backbones to act as a hub.

2.18: AS Core Internet graph

chief cartographers: Brad Huffaker, Andre Broido, K. Claffy, Marina Fomenkov, Sean McCreary, David Moore, and Oliver Jakubiec (Cooperative Association for Internet Data Analysis – CAIDA).
aim: to visualize the macroscopic structure of the Internet for a snapshot in time by showing relationships between peer and geographic location of ISPs.
form: polar-projected graph.
technique: data gathered from automatic measurements of the Internet topology through a program skitter.
date: January 2000.
further information: <http://www.caida.org/analysis/topology/as_core_network/>

The evocative images shown here look like strange nebulous jellyfish with complex dense fronds, floating in a black sea. In fact, they are the preliminary results from a visualization of Internet topology of connections between routing traffic nodes. The visualizations use a special type of 3-D projection called hyperbolic space to enable the interactive display of huge directed graphs, some with hundreds of thousands of nodes. They are produced using a software tool code-named Walrus currently being developed by CAIDA researcher Young Hyun. The tool works interactively, so that the user can rotate and view the graph from any position.

Walrus uses hyperbolic space so that a large graph can be displayed in the limited resolution of a computer screen and still be legible to the analyst. Hyperbolic projection provides a "focus + context" view of data, whereby the viewer sees the world through a 3-D fish-eye lens so that the portion of the graph at the center of the display is greatly enlarged, revealing fine detail. Despite this focus, the rest of the graph is still visible, gradually diminishing in size (but never disappearing) towards the edge of the fish-eye, and so providing vital overall context. The user can drag any portion of the graph into the center of the fish-eye and it will enlarge in a smooth transition.

(The Walrus software is based in part on the 3-D hyperbolic research of Tamara Munzner, which we examine in chapter 3.)

The data being visualized by Walrus is gathered by skitter, a comprehensive Internet-monitoring tool developed also by the CAIDA researchers. skitter's monitoring software runs on dedicated machines located at a growing number of sites on the Internet, and it actively probes tens of thousands of destination points on a daily basis, recording the route and performance of test traffic. The three images show the routing topology of the Internet as measured by the skitter monitor based in London, visualizing the same dataset in different ways for some 535,000 separate nodes and over 600,000 links.

2.19: 3-D hyperbolic visualization of Internet topologies

chief cartographer: Young Hyun (Cooperative Association for Internet Data Analysis – CAIDA).
aim: to provide interactive exploration of huge graphs (greater than 100,000 nodes) showing complex structures of Internet routing.
form: dense and organic, jellyfish-like 3-D graphs projected inside a transparent sphere.
technique: data gathered from automatic measurements of the Internet topology by skitter. Visualization through custom-written hyperbolic graph-viewer code named Walrus.
date: 2000.
further information: <http://www.caida.org/~youngh/walrus/walrus.html >

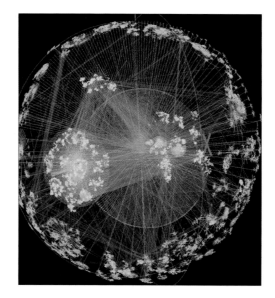

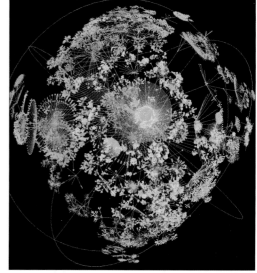

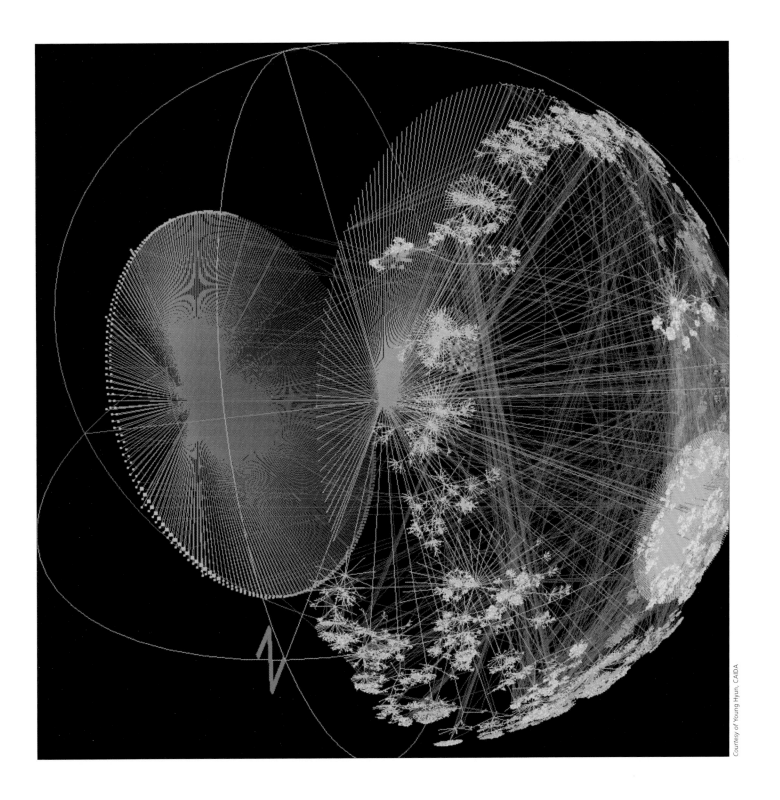

The top image opposite is a striking 3-D topology spatialization of vBNS, the high-performance Internet network that links US universities and research labs (http://www.vbns.net). It is part of a short animation, created by the Cichlid visualization toolkit, that simulates data packets zooming along the links as dark-red cubes. The topological structure – copied from a more conventional-looking static 2-D spatialization shown below – uses a standard arc–node graphical form.

The core vBNS network hubs are represented by the large blue cubes interconnected by purple arcs. These are laid out in space based on approximate geographic configuration, and so west-coast nodes are shown on the left-hand side of the image and east-coast nodes are on the right. Branches from these blue cube hubs are links to aggregation nodes (shown by cyan pyramid symbols) and normal university nodes (small red balls). Also shown are connections to other Internet networks represented by larger gray balls. All nodes are labeled with name (or abbreviation) of the site or network. The overall image is a still frame from an animation of simulated Internet traffic flows.

2.20: 3-D map of the topology of the vBNS Internet backbone

chief cartographer: Jeff Brown (Measurement & Operations Analysis Team, National Laboratory for Applied Network Research).
aim: to demonstrate the potential of the Cichlid visualization toolkit to model networks. This example models the topological structure of the vBNS network in the United States in 3-D with an animation of simulated data packets.
form: 3-D arc–node network representation floating in empty space. Layout of the graph is partially geographic. Data packets are shown by small red blocks.
technique: digital animations and images created by Jeff Brown's Cichlid visualization toolkit.
date: summer 1999.
further information: MOAT homepage at <http://moat.nlanr.net> Cichlid homepage at <http://moat.nlanr.net/Software/Cichlid/>
further reading: "Network Performance Visualization: Insight Through Animation", by Jeff Brown, A. J. McGregor and H-W. Braun, paper for PAM 2000 Workshop, Hamilton, New Zealand, April 2000. <http://moat.nlanr.net/~jabrown/cichlid-misc/cichlid-pam2k.pdf>

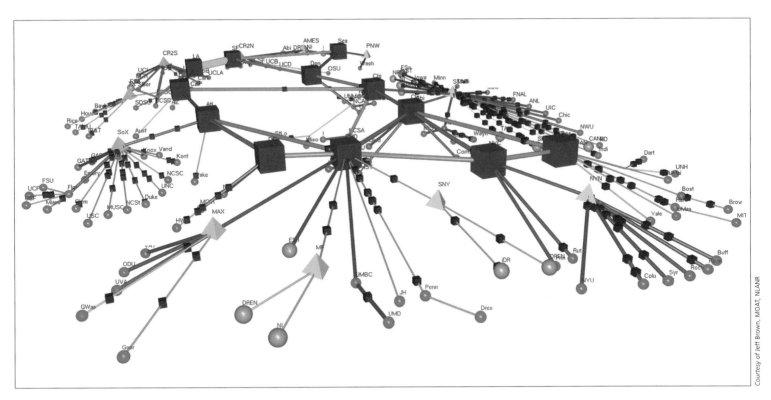

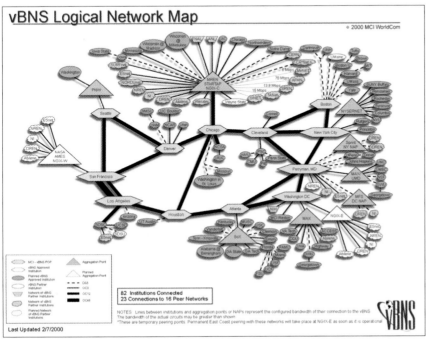

The geography of data flows

Whilst the maps and spatializations discussed so far reveal information concerning network topologies, they tell us very little about how much data actually flow through those networks or what the data consist of. There are considerable difficulties in obtaining representative data on traffic flows across the Internet. It is also a challenge to represent what are temporal phenomena, and so cartographers often resort to various aggregations over time and space. In this section we examine several different attempts to produce maps that perform this role.

The maps shown opposite are some of the earliest of Internet traffic flows still publicly available. They display data traffic flows on the Usenet network in 1993 (see page 164 for a fuller discussion of Usenet). They were created as part of a project to measure the number of hosts in the Usenet network and the amount of traffic (number and volume of posted articles) flowing between them, along with estimates of the number of readers, from 1986 to the early 1990s. In order to visualize the geography of the Usenet network infrastructure and traffic flow within this network, a mapping tool called Netmap was developed. The maps that Netmap produced were somewhat rudimentary in style, comprising a simple arc–node network representation plotted on a simple black-and-white world map with national boundaries shown, where the width of the arcs was proportional to volume of traffic flowing over the links between backbone nodes.

This mode of representation has led to severe problems of graphical overplotting in North America and Europe, where most Usenet sites are based and through which most traffic flows (see top image). In fact, the map reveals that very little of the rest of the world was connected to the Usenet system in 1993. The problem of overplotting is partially remedied by showing only the data flows between backbone sites and by producing region-specific maps. The backbone nodes are represented by labeled circle symbols giving the ID code for the site. The more numerous ordinary sites on the Usenet network are shown by small black dots.

2.21: Usenet data flows

chief cartographer: Brian Reid (while at the DEC Western Research Labs, USA).
aim: to display worldwide Usenet data flows between individual sites.
form: black-and-white line world map of countries, overlaid with a conventional arc–node network representation, with line width proportional to data-flow volume.
technique: created with Reid's Netmap software and output as a Postscript digital map.
date: May 1993.
further information: a series of Reid's maps is available in Postscript format at <ftp://gatekeeper.dec.com/pub/maps/>; Reid's homepage is at <http://www.reid.org>

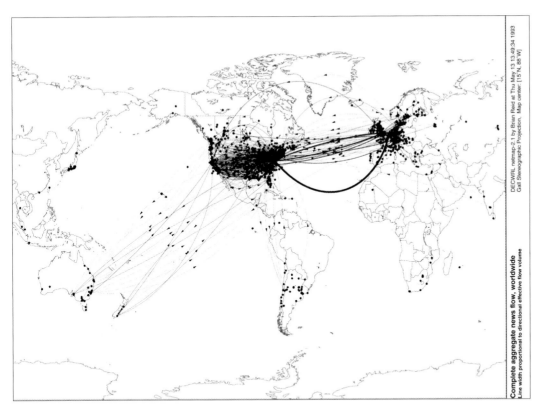

Complete aggregate news flow, worldwide
Line width proportional to directional effective flow volume

DECWRL netmap-2.1 by Brian Reid at Thu May 13 13:49:34 1993
Gall Stereographic Projection, Map center: [15 N, 88 W]

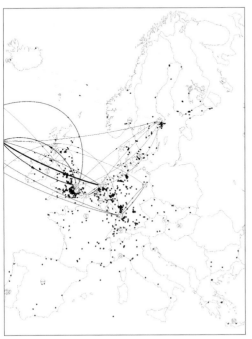

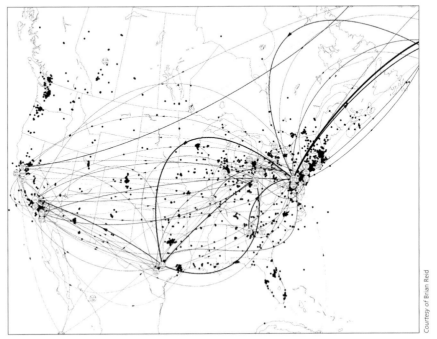

TeleGeography's maps develop further the cartographic style employed in Reid's Usenet maps. The maps display aggregate telecommunication traffic-flows across the public telephone network, as measured in millions of minutes for 1999, for two important regions of the world: Europe (opposite) and Asia (below). The nodes and links are used to represent two aspects of the data. The nodes, positioned on the approximate geographic location of the capital city of each country, are drawn as circles with their diameter proportional to the annual total volume of outgoing traffic for that country. The traffic flows between countries are represented by lines sized proportionally to the total annual traffic between that pair of nodes. Only links over a minimum threshold are shown, so as to limit the problems of overplotting.

Despite the fact that the data flows are not normalized in relation to population, the maps reveal a great deal about communications and data flows between countries in these regions. Some large flows are due to business transactions – for example, those at the heart of Europe – whereas others mainly represent large groups of immigrants keeping in contact with home, such as the large flows between the United Kingdom and Ireland, and Saudi Arabia and India.

2.22: TeleGeography's telecommunications flow maps

chief cartographer: Gregory C. Staple and colleagues (TeleGeography, Inc.).
aim: to map the volume of telecommunication traffic between countries.
form: map of countries, with traffic flows represented by smoothly curving lines, and nodes as circular symbols plotted onto capital-city locations. The size of the circular node indicates the country's total outgoing traffic, while arc thickness is proportional to the traffic on that route.
technique: hand-crafted maps using arc–node representation.
date: 1999.
further information: TeleGeography's homepage at <http://www.telegeography.com>

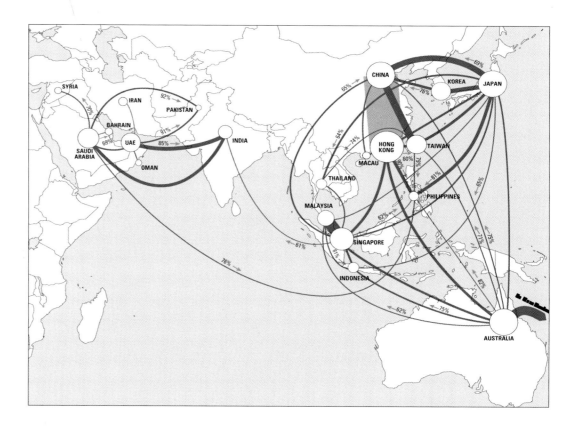

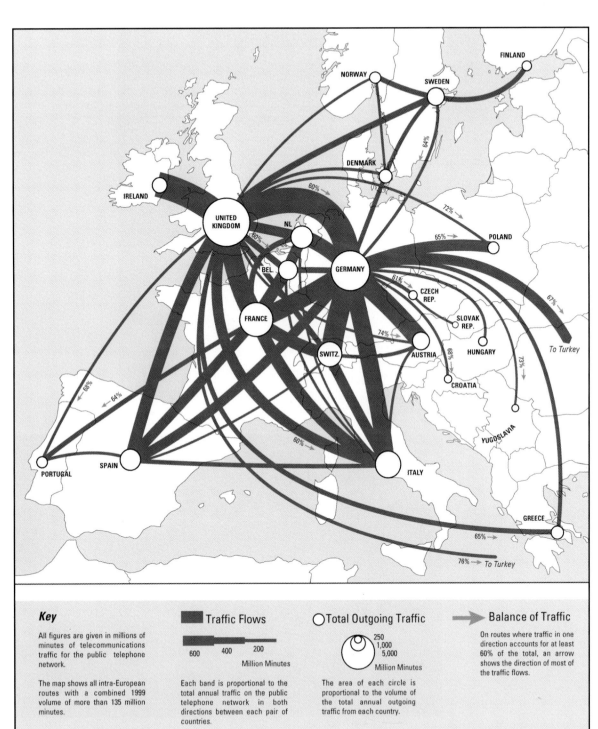

Key

All figures are given in millions of minutes of telecommunications traffic for the public telephone network.

The map shows all intra-European routes with a combined 1999 volume of more than 135 million minutes.

Traffic Flows

600 400 200

Million Minutes

Each band is proportional to the total annual traffic on the public telephone network in both directions between each pair of countries.

Total Outgoing Traffic

250
1,000
5,000

Million Minutes

The area of each circle is proportional to the volume of the total annual outgoing traffic from each country.

Balance of Traffic

On routes where traffic in one direction accounts for at least 60% of the total, an arrow shows the direction of most of the traffic flows.

Countries shown on map: FINLAND, NORWAY, SWEDEN, DENMARK, IRELAND, UNITED KINGDOM, NL, POLAND, BEL., GERMANY, CZECH REP., SLOVAK REP., FRANCE, HUNGARY, AUSTRIA, SWITZ., CROATIA, YUGOSLAVIA, PORTUGAL, SPAIN, ITALY, GREECE, To Turkey

Percentages on map: 64%, 60%, 72%, 65%, 61%, 67%, 74%, 68%, 73%, 69%, 64%, 60%, 65%, 76%

Courtesy of TeleGeography, Inc.

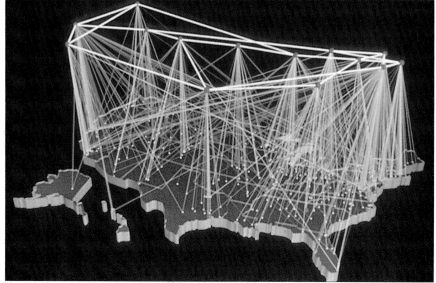

The two striking images shown opposite are from one of the most well-known Internet visualizations. They are single frames from an animation of Internet traffic flows on NSFNET for the early 1990s. NSFNET was the heart of the Internet from the late 1980s to 1995, connecting major universities in the United States and linking them with many countries around the world. Here, a visual metaphor of a virtual network floating above the United States is used to represent the NSFNET backbone, with the source of the inbound data and its connecting node shown by the colored vertical lines. These lines are color-coded to indicate the volume of traffic that was carried, ranging from low (purple) to high (white).

The bottom map shows the "T1" backbone of NSFNET, summarizing inbound traffic flows for September 1991 (all quantities hereafter representing the aggregate volume for that month). A stylized base map of the US states provides a context for the virtual network. The white links are carrying 100 million bytes. The top map shows the next generation of the network, the T3 ANS/NSFNET, against a more "realistic" background of the Earth's surface taken from a satellite terrain image. The color-coded lines projecting from the ground represent the network links from individual sites (mostly universities) to the backbone. The scale of data flows represented by colored links increased in magnitude during the intervening period to December 1994, and so white now represents 1 trillion bytes.

The maps in combination reveal that the dominant traffic flows are on the US east and west coasts where most of the population and universities are located, and that there was significant growth in the number of sites connected to the network between 1991 and 1994.

2.23: Visualizations of traffic flows over NSFNET

chief cartographers: Donna Cox and Robert Patterson (National Center for Supercomputing Applications, University of Illinois Urbana-Champaign, USA).
aim: to visualize the growth in traffic flows on the NSFNET backbone network in the United States in the early 1990s.
form: a virtual backbone floating in space above the United States, with lines color-coded according to the volume of the traffic carried.
technique: high-quality 3-D animation using custom software.
dates: visualizing data for (a) September 1991 and (b) December 1994.
further information:
<http://www.ncsa.uiuc.edu/SCMS/DigLib/text/technology/Visualization-Study-NSFNET-Cox.html >

One of the urban myths concerning the origins of the Internet was that it was designed to survive a Soviet nuclear strike through its distributed structure. The map opposite is reminiscent of this myth, having something of the appearance of ballistic missile tracks displayed on a NORAD (North American Aerospace Defense) radar display. The world is encompassed by a fountain of arcing trails streaking from country to country. In fact, this impressive "arc map" is a visualization of Internet traffic flows between 50 nations, as measured by the NSFNET backbone for a two-hour period in February 1993; it is one of the most compelling, almost iconic, representations of the Net, and it is widely reproduced in newspapers and magazines and even on book covers.

The color, thickness and height of the arcs are used to encode the traffic statistics for particular international routes. The arcs are also partially translucent so as not to completely obscure lines at the back of the map, while their height above the base map is in relation to total volume of traffic flowing over a link. This has the effect of making the links with the fastest flow the highest and therefore most visually prominent on the map.

The user had considerable interactive control over the map within the SeeNet3D application. For example, the arc-height scaling and translucency could be smoothly varied; and the map could be rotated and scaled, so that the user could view it from any angle.

The map shows that there was significant traffic, in the early 1990s, between three areas of the world: North America and Europe; Europe and Australasia; and Australasia and North America. Most traffic crossed the Atlantic. The map does not show all traffic, however, because it is limited to just 50 countries. As such, it portrays a selected image – one that is dominated by anglophone countries, which were the principal nations connected to the Internet in 1993.

2.24: Internet traffic flows between countries
chief cartographers: Stephen Eick and colleagues Ken Cox, Taosong He and Graham Wills (Bell Labs-Lucent Technologies / Visual Insight).
aim: to create compelling 2-D and 3-D visualizations to understand network data flows.
form: flat world map with country boundaries and traffic flows represented as looping arcs traversing the Earth.
technique: a still image from an interactive visualization tool called SeeNet3D, designed for network visualization and data exploration.
date: 1996.
further information: Eick's Network Visualization Gallery
<http://www.bell-labs.com/user/eick/NetworkVis.html>
further reading: "3D Geographic Network Displays", by Kenneth C. Cox, Stephen G. Eick, and Taosong He, ACM Sigmod Record, 25(4), pp. 50–54, December 1996.
<http://www.bell-labs.com/user/eick/bibliography/1996/3D_copyright.pdf>

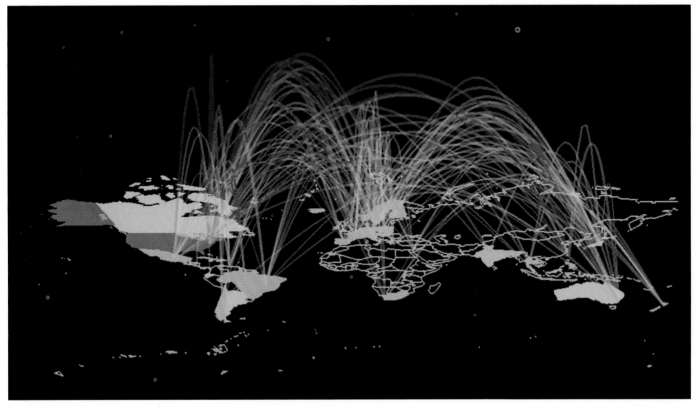

Courtesy of Stephen Eick, Bell Labs–Lucent Technologies / Visual Insight

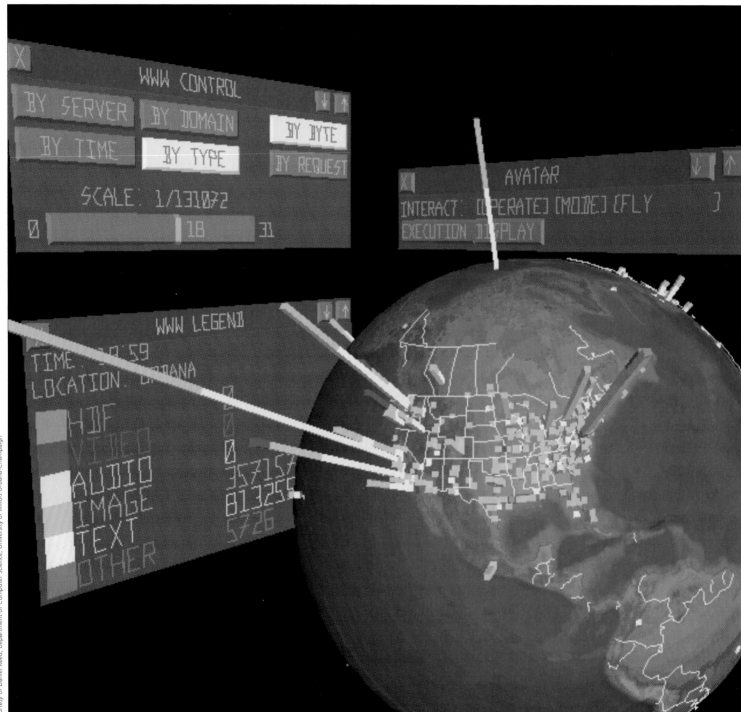

The images here display a moment in the dynamic geographic visualization of Internet traffic. The maps were one outcome of a larger project to monitor and analyze, in real time, the patterns of traffic on Web servers to help design systems that perform better. The visualizations utilize what might be termed a "skyscraper" metaphor mapped onto the globe to reveal Web traffic and type, originating at different locations in real time. As such, the heights of the skyscrapers grow and shrink as the traffic flows vary. The skyscrapers are located in the geographic location of traffic origin and are divided into segments, with each segment representing different types of Web traffic (text, images, video, etc.) that make up the total volume.

The two figures show Web traffic in North America, although Europe can clearly be seen on the horizon. The data are Web requests to the National Center for Supercomputing Applications (NCSA) Web server in 1995. This server was the home of the Mosaic Web browser, at that time one of the most popular sites on the Web and receiving over 400,000 "hits" (requests for data) on an average day. As hits were received, the display was updated. The system could cope with a peak traffic load of 50 hits per second. To enable the visualization of traffic on the globe, the cartographers developed a method for determining the geographic location of the origin of Web traffic. However, this only produced approximate locations, with traffic originating outside the United States simply being aggregated to national level, with the capital city assigned as the location of origin. Within the United States, much greater spatial discrimination was achieved, with traffic being mapped to the city of origin.

It is important to note that the global "skyscraper" maps were designed to be accessed in a sophisticated virtual reality (VR) system that allowed users to immerse themselves in the visualization. This required expensive virtual-reality hardware, including head-mounted displays and even a CAVE. (A CAVE is an expensive state-of-the-art virtual-reality environment that uses a room-sized cube of back-projected screens to display the scene. Users can walk around in the CAVE, and the scene is updated in real-time response to their movements. In these VR-supported environments, the user's head movements are continuously tracked and the view constantly updated in response to where the user is looking.) Moreover, it was possible to interact with the map by reaching out and "touching" one of the skyscraper bars to initiate an enquiry and see full details of traffic flows (opposite). Control panels and a map key were also displayed. The user had considerable power to change the data classification and the scale and rotation of the globe.

2.25: Geographic visualization of traffic on the NCSA website

chief cartographers: Stephen E. Lamm, Daniel A. Reed and Will H. Scullin.
aim: to show geographic origin of traffic, in real-time, of a popular website.
form: 3-D globe, with traffic represented as bars projecting from the Earth's surface looking like virtual skyscrapers. Height, location, and color-coding of the bars encode data attributes for the website traffic.
technique: custom software to analyze traffic and create 3-D visualization using virtual-reality techniques.
date: August 1995.
further information: "Real-Time Geographic Visualization of World Wide Web Traffic", by Stephen E. Lamm, Daniel A. Reed, and Will H. Scullin, proceedings of the Fifth International World Wide Web Conference, 6–10 May 1996, Paris.
<http://www5conf.inria.fr/fich_html/papers/P49/Overview.html>

Mapping traceroutes

Given the rapidly evolving nature of the Internet, maps employing static data (all those so far described) quickly become dated. To counteract this limitation, research is being undertaken to create dynamic maps that are automatically updated in real time. One way that this can be achieved is through the mapping of traceroutes. A traceroute is an Internet utility that reports the route that data packets travel through the Internet to reach a given destination, and also the length of time taken to travel between all the nodes along a route. The original traceroute tool was created by Van Jacobson, a computer scientist working at Lawrence Berkeley national laboratory in the United States in 1988.

Traceroutes are an important Internet debugging tool for those involved in keeping networks running. They can help identify routing problems quickly and simply; moreover, they can be useful for tracking down a source of spam and also for finding out where a website is located before credit-card details are revealed. On another level, traceroutes can help to satisfy people's curiosity to know how their computer connects to the global Internet and how they can access information from all around the world as if it were just next door. Traceroutes reveal the hidden complexity of data flows, traversing many nodes – usually owned and operated by competing companies – to reach a given destination.

A typical output of a basic traceroute utility is shown here, charting the route between Maynooth in Ireland to Adelaide in Australia, via New York and San Francisco.

This approach has now been complemented by a new breed of geographic and visual traceroutes, and we look next at one good example: VisualRoute.

1	2 ms	4 ms	2 ms	cismay.may.ie [149.157.1.6]
2	4 ms	5 ms	4 ms	Moskva-s2-2-nuim.hea.net [193.1.194.13]
3	6 ms	6 ms	5 ms	Kinnegad-fe0-0.hea.net [193.1.198.253]
4	130 ms	98 ms	96 ms	heanet.ny3.ny.dante.net [212.1.200.101]
5	103 ms	96 ms	106 ms	[212.1.201.36]
6	101 ms	102 ms	106 ms	500.POS2-0.GW6.NYC9.ALTER.NET [157.130.254.245]
7	97 ms	112 ms	97 ms	527.at-5-0-0.XR1.NYC9.ALTER.NET [152.63.24.66]
8	99 ms	98 ms	97 ms	181.ATM5-0.BR3.NYC9.ALTER.NET [152.63.23.145]
9	103 ms	99 ms	98 ms	acr2-serial3-0-0-0.NewYorknyr.cw.net [206.24.193.245]
10	*	177 ms	172 ms	acr2-loopback.SanFranciscosfd.cw.net [206.24.210.62]
11	178 ms	172 ms	178 ms	optus-networks.SanFranciscosfd.cw.net [206.24.209.206]
12	489 ms	488 ms	487 ms	[192.65.89.65]
13	486 ms	484 ms	486 ms	GigEth0-0-0.ia4.optus.net.au [202.139.190.18]
14	512 ms	514 ms	522 ms	SA-RNO-Int.ia4.optus.net.au [202.139.32.206]
15	523 ms	517 ms	510 ms	lis255.atm1-0.main.flinders.gw.saard.net [203.21.37.4]
16	513 ms	514 ms	515 ms	rory.cc.flinders.edu.au [129.96.253.11]

Trace complete

VisualRoute presents traceroute results in both a table form and plotted onto a world map. In this case, the routes from London to three distant websites are shown – first to the Japanese Prime Minister's Office (www.sorifu.go.jp), then to the Russian Parliament (www.duma.ru) and finally to the White House website (www.whitehouse.gov).

Each row in the table within the illustration represents one leg in the route. The columns in the table provide much useful information, such as the domain name of the machine at each link-point, its approximate geographic location, and the company name of the networks that are being traversed. The round-trip time for each leg of the journey is also shown, as both a numeric value and also on a graph. In all of our three examples it takes more than 15 hops, crossing multiple networks – quite a feat of routing and cooperation that happens unseen to the millions of average Web surfers.

A limited amount of user interaction with the map is possible through panning and zooming. VisualRoute is limited by its geographic database of node locations, although it allows users to add to this database and to add their own, more detailed, base maps. One of the significant limits of conventional

traceroute utilities such as VisualRoute for dynamically exploring the Internet's infrastructure is that the origin of the trace is fixed to one location, namely the computer on which the trace is run. This limitation of a fixed point of origin has been partially overcome by the development of Web-based traceroutes.

There are also difficulties in mapping an Internet node (logically identified by a network address) to an actual latitude and longitude on the globe. There is no automatic means to achieve this mapping of virtual addresses to real-world addresses. Consequently, these programs use a variety of heuristics to try to resolve a network node to a geographic location with differing levels of accuracy. This is a difficult problem to solve and each utility experiences varying degrees of success in determining a location. Ultimately, they are dependent largely on looking up the Internet addresses in static databases of latitude and longitude, but these are partial and cannot keep pace with the dynamic growth and change of the global Internet.

2.26: VisualRoute geographic traceroute

chief cartographer: Jerry Jongerius (FORTEL).
aim: to trace how data flows through the Internet in real time.
form: the trace shown as a black line, plotted on a base map.
technique: simple plotting of trace onto base map.
date: initially released in January 1998 (screenshots taken in November 2000).
further information: VisualRoute homepage <http://www.visualroute.com>
further reading: *Mapping the Internet with Traceroute*, by Jack Rickard, Boardwatch Magazine, December 1996. <http://boardwatch.internet.com/mag/96/dec/bwm38.html>
Nailing Down Your Backbone: The Imprecise Art of Tracerouting, by Jeffrey Carl, Boardwatch ISP Directory, Summer 1999. <http://www.ispworld.com/isp/traceroute_art.htm>

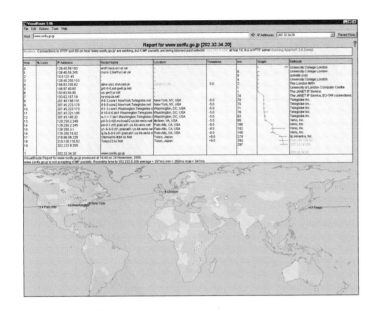

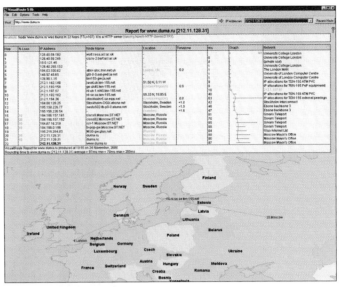

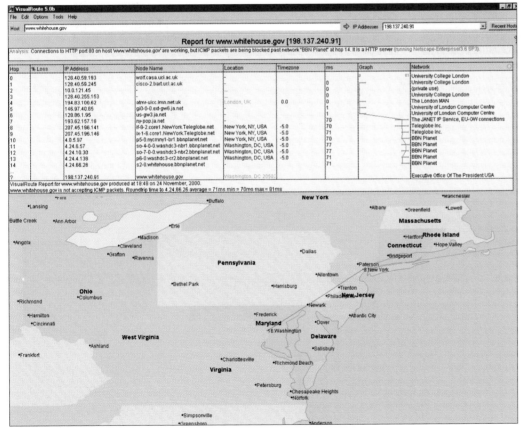

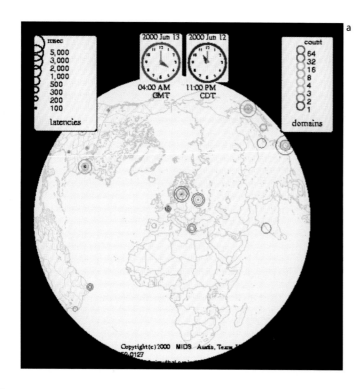

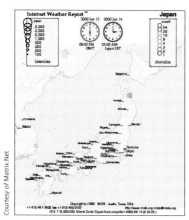
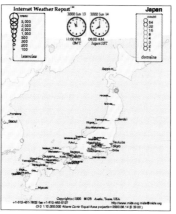
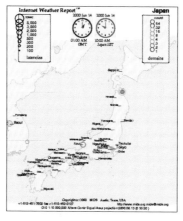

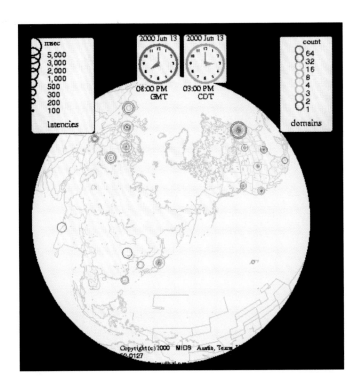

What's the Net "weather" like today?

Like other types of network, such as a national road system, the Internet can experience congestion. One specialized type of data-traffic map seeks to chart the flow of data throughout a network, identifying how long it takes for data to travel through the system. This type of data can be a valuable aid for network administrators seeking to keep their systems free of hold-ups. Here, we detail two example mappings.

The performance of the global Internet is measured every four hours, every day of the year, by Matrix.Net to produce a unique and continuous "Internet Weather Report" (IWR). Having been running since 1993, IWR gives one of the few consistent, time-series measurement bases of the global Internet. Frames are shown from IWRs at the global scale and for Japan from June 2000.

Forecasts are made six times a day for over 4,000 Internet sample points all around the world. This measurement consists of "pinging" (sending a tiny packet of data to) sample computers from Matrix.Net's headquarters in Austin, Texas, and measuring the time the ping data takes to travel there and back. This round-trip time, measured in milliseconds, gives an estimate of the latency for that sample point on the Internet. These ping measurements are turned into a map with graduated circle symbols to represent latency (the larger the circle, the longer the delay). The color of the circle represents the number of Internet hosts at each location that are being surveyed by the IWR. Individual maps are used to create daily animations of traffic congestion on the Internet. Matrix.Net can produce these maps from a global to a local scale.

In basic terms, small circles on the map show a healthy Internet, while large circles are indicative of poor performance and possible problems. Interpretation of IWR animations can be difficult, although it is claimed that you can see patterns of Internet latency ("storms" of congestion). A major limitation with IWR is that all the measurements are taken from a single point of origin at the company's headquarters, which means that each map can never give a representative view of the whole of the Internet's weather. To overcome this limitation, Matrix.Net has developed a much more comprehensive monitoring system called Internet Average (see <http://www.miq.net/>), which samples Internet performance continuously from many different points across the world.

2.27: Internet Weather Report (IWR)

chief cartographers: John Quarterman and colleagues (Matrix.Net).
aim: to show the Internet "weather" (performance/congestion) at hundreds of locations across the world.
form: animated geographic maps, with graduated color-coded circles showing changing latency patterns.
technique: measurement six times a day from the Matrix.Net company base in Austin, Texas. Custom software generates daily animated maps for different regions and countries.
date: examples taken from June 2000. IWR has been in operation since 1993.
further information: see <http://www.matrix.net/isr/weather/>

The diagrams shown opposite display the daily peak traffic for individual network links on NORDUnet, the education-and-research network for the Nordic region, with the map below displaying the countries connected and links to the rest of the Internet. The top spatialization opposite provides a detailed view of the topology of nodes and links in the network. The bandwidth of individual links is represented by line thickness. The color-coding of the lines denotes the percentage peak traffic-load for the day, in 10-percent classes, as specified in the lower-left legend. Thus dark-blue links represent a peak traffic-load of between 0 and 10 percent, while the most congested links, colored dark red, peak at 90–100 percent traffic load. The links are divided into two segments, with the segment closest to a node (shown as a box) representing outgoing traffic for that node.

The spatialization is a summary of, and visual interface to, a complex measurement system that monitors the network traffic in great detail. More detailed statistics can easily be obtained by clicking on individual links of interest. The statistics – for example average and peak traffic on an hourly basis – are then presented on standard line charts. This type of analysis allows system administrators to carefully monitor their system's use and to counter any particular difficulties such as little spare capacity.

2.28: NORDUnet network-traffic map

chief cartographer: Rami Heinisuo (NORDUnet).
aim: to show the peak daily percentage traffic load on individual links in the network. Also, to provide an index to more detailed traffic statistics.
form: topological map of the network, with links color-coded by density of traffic. Nodes are shown by the labeled squares.
technique: custom-written software (NORDstat) for network measurement and the creation of the interactive traffic map on a daily basis.
date: example map from 13 June 2000.
further information: NORDUnet homepage at <http://www.nordu.net> traffic map at <http://www.nordu.net/stat-q/load-map/ndn-map,,traffic,busy>

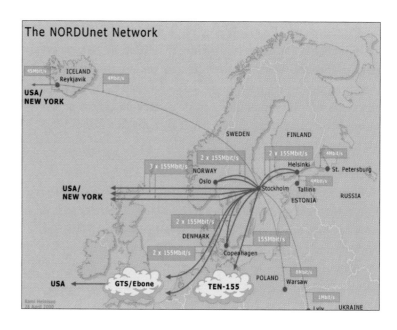

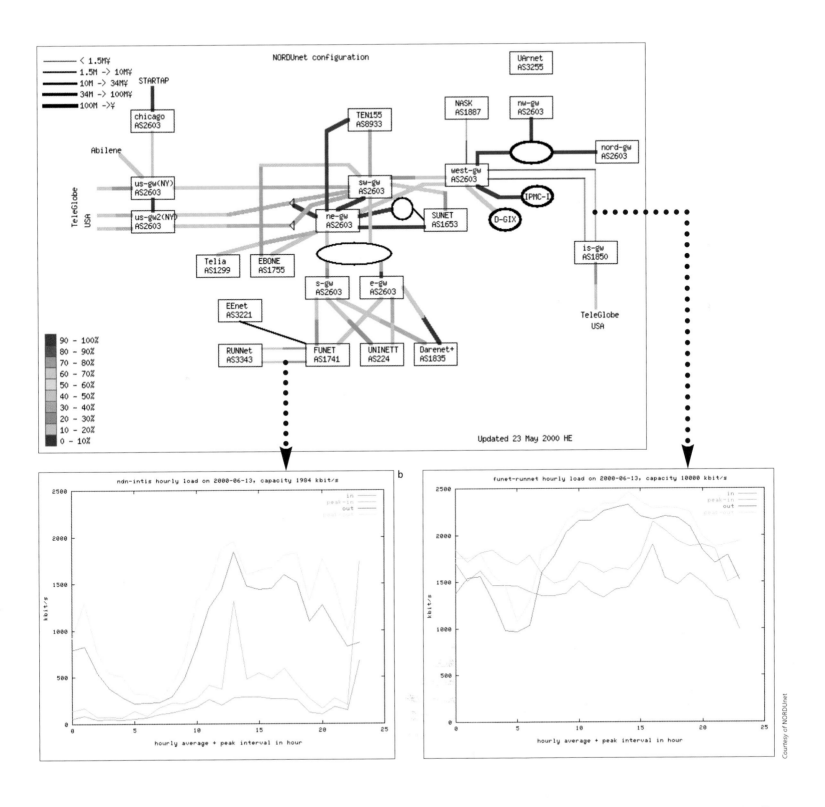

Mapping cyberspace usage in temporal space

As the "Internet Weather Reports" (page 66) demonstrate, there is a temporal aspect to the use of the Internet, not only in terms of daily fluctuations in Web usage but also in relation to delays in the transfer of traffic. Analysis of these "weather" maps reveals that peaks in congestion coincide with the overlapping of key times in geographic space – for example, the coinciding of early morning on the east coast of the United States with mid-afternoon in Europe, or the coinciding of early morning on the US west coast with early afternoon in the east coast. In this final section of chapter 2, we detail an attempt to map cyberspace usage at a global scale in relation to time.

The innovative and unconventional-looking map (opposite) shows the number of people who could potentially be networked in cyberspace during office hours (9a.m.–5p.m.) in terms of land-line and mobile telephones and the Internet throughout the 24 hours of a single day. The map uses the visual style of a radar graph from statistical graphics, with three different color-coded polygons representing the different communications technologies. The size of the polygons varies through the hours of the day, with the largest percentage of people online being shown when the polygon vertices stretch furthest from the center. Time around the edge of the map is calibrated to London (GMT), with three office days (Shanghai, Paris and New York) highlighted by the circular strips surrounding the map. The area enclosed by the red line represents the percentage of Internet hosts operational during office hours. This peaks in the afternoon, London time, as the Europeans are still in the office and the North Americans are just starting work. The green- and orange-lined polygons represent fixed and cellular telephone connections, which peak at two different periods – first in the morning, London time, when Europe goes to work and Asia is still in the office, and then in the afternoon when the North Americans and Europeans are together in network time–space. Interestingly, in the first European morning peak the percentage of Internet hosts is relatively low (at around 30 percent) compared with the number of people connected by telephone; this reflects the dominance of the United States on the Internet.

The world's population is shown by the blue polygon and its particular temporal pattern makes an interesting contrast, in that for part of the day it is outside the other three polygons. This means that many more people could be online (i.e. they are awake during their office hours of 9–5) but are not connected to any global communications networks by the telephone or Internet. This is most apparent through the early morning, London time (3–8 p.m.), when Asia is awake but the highly networked nations in Europe and North America are asleep. At the peak of the networked population in the afternoon, London time, the blue line is close to the center of the map, showing that this represents less than half the world's population. This demonstrates graphically that cyberspace is accessible to the privileged minority of the world's population, located predominantly in the wealthy Western nations. Interpreting their creation, TeleGeography concludes in a 1999 report that: "Each day, while the demand for connections moves clockwise from East to West with the sun, available network capacity appears to move counterclockwise as network resources are idled during the night."

2.29: Circadian geography of communications networks
chief cartographers: Zachary Schrage and Gregory Staple (TeleGeography, Inc., Washington D.C.).
aim: to show how many people are connected via the Internet and via fixed and mobile telephones around the world through 24 hours of a day.
form: radial graph on a polar projection.
technique: hand-crafted map.
date: 1998.
further information: TeleGeography homepage at <http://www.telegeography.com>
further reading: *TeleGeography 1999: Global Telecommunications Traffic Statistics and Commentary*, edited by Gregory C. Staple, 1998, TeleGeography, Inc., Washington, DC.

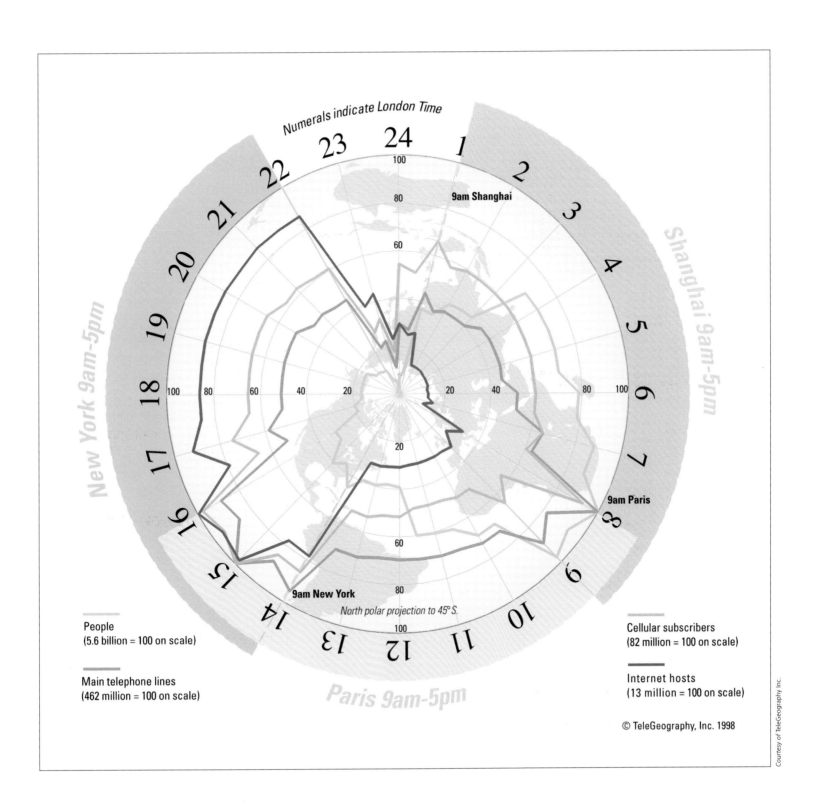

Numerals indicate London Time

9am Shanghai

9am Paris

9am New York

North polar projection to 45° S.

New York 9am-5pm

Shanghai 9am-5pm

Paris 9am-5pm

People
(5.6 billion = 100 on scale)

Main telephone lines
(462 million = 100 on scale)

Cellular subscribers
(82 million = 100 on scale)

Internet hosts
(13 million = 100 on scale)

© TeleGeography, Inc. 1998

chapter 3

Mapping the Web

Whilst maps of infrastructure and traffic are the most commonly produced spatial visualizations of cyberspace, perhaps the most exciting projects are occurring in relation to mapping the informational landscape of cyberspace. Of all the types of mapping and spatialization we present, this is clearly the area where the most original research and development is taking place. Here, researchers from the disciplines of computer graphics, information design, human–computer interaction, virtual reality, information retrieval and scientific visualization, along with those working for commercial enterprises, are seeking ways in which to spatialize cyberspace itself, trying to find ways to:

■ improve modes of navigation through, and searching of, the information spaces of the Internet;

■ provide media that are easier to comprehend;

■ document the extent and territories of different media.

In other words, different methods of spatialization – where a spatial structure is applied to data where no inherent or obvious one already exists – are being experimented with and developed in order to try to enhance the usability and comprehensibility of cyberspace, exploiting the fact that people find it easier to process and understand visual displays than large volumes of written text or columns of numbers.

Many of these spatialization techniques are being developed for future commercial exploitation, as it is believed that the first successful software will become a "killer" (i.e. pervasive and dominant) application, replacing hypertext-driven browsers with a hyper-visual mode of navigation. The reason why such a "killer" application has not been successfully developed or marketed to date is because, as noted in brief in chapter 1, developing effective spatializations is not an easy task. Cyberspace, although not spaceless as hypothesized by some analysts, does not have a static, easily measurable space–time geometry. This is because cyberspace is entirely socially produced – its form and structure are dematerialized and determined by its creators. Space–time, then, is discontinuous

and relative and changes from media to media and site to site. Moreover, cyberspace is highly mutable – able to change form yet still possess the same content – and dynamic, with new material being constantly added and old material altered, updated and deleted, leaving no record. These changes are often "hidden" until encountered. As time unfolds, the mutability and dynamism of cyberspace will increase accordingly, so that its contents are continually evolving, disappearing and restructuring at ever increasing rates. Effective spatializations, especially for navigation and search, must be able to cope with such changes, automatically and seamlessly incorporating them. Furthermore, spatializations, if they are to be used as modes of navigation and query, need to be interactive, so that territory and map become one.

In the rest of this chapter we detail the methods so far developed to try to spatialize cyberspace and discuss some of the spatializations produced by these techniques. We focus our attention on spatializations of information, examining attempts to spatialize the social domains of cyberspace (e.g. email, chat, multi-user domains, virtual worlds, games) in the following chapter. In order to provide a structure to our discussion, we have sequenced our examples of maps along a scale continuum, starting with individual websites and progressing through to large sections of the Internet.

Information spaces of the Internet

A useful place to start this chapter is to examine the conceptual map of cyberspace created by John December in 1994. This spatialization provides a good overview of the conceptual spaces of the Internet at that time. It illustrates that the Internet supported a range of interconnected information spaces beyond the Web, each of which has differing virtual "geographies" (also see chapter 4).

John December is a successful Internet consultant running his own firm, December Communications, Inc., based in Milwaukee, Wisconsin. December has written and presented extensively about the Internet, the Web and broader issues of computer-mediated communications. He created a number of different conceptual cybermaps in the mid-1990s to try to give a "big picture" overview of the nature of cyberspace. The conceptual diagram shown opposite reveals the geographies of net spaces, providing a good way of conceptualizing them as distinct and self-contained domains, but with fluid and irregular boundaries and many interconnections and overlaps between them. The map also illustrates that other networks existed beyond the Internet, such as FidoNet and BITNET, that make up the larger, globe-spanning Matrix of computer-communications systems (see chapter 2).

December's map is hand-drawn and has a sketch-like quality. The networks and information spaces are drawn as puddle-like "blobs". There are several distinct networks, labeled in purple, showing the diversity of cyberspace in 1994. "The Matrix" is the largest "blob" and contains the Internet (and its information spaces running over TCP/IP) as the largest network, along with the notable, but much smaller, networks of FidoNet, UUCP, and BITNET. There was also a sizeable presence of commercial online services in the early 1990s (such as AOL, CompuServe, Prodigy, and GEnie) which were generally closed and proprietary in nature, and not really part of the Internet.

The heart of the spatialization focusses on the eight most significant information spaces in the mid-1990s as identified by December. These were the Web, Gopher, email, finger, WAIS, telnet, ftp and Usenet. Most of these information spaces are contained within the Internet, except for email and Usenet which spill over to other networks. The key landmarks for each information space are clearly labeled and the connections between spaces are represented by red arrows. So, for email space in 1994, one of its landmarks was the Publicly Accessible Mailing Lists (PAML) index, while the Web's landmarks were the then new search engines and directories such as Lycos, Web Crawler, Yahoo! and CERN's venerable Virtual Library.

What makes these information spaces different from one another are the modes of information exchange, degrees of synchronicity and levels of social interaction they support. At a fundamental level, the differences between information spaces are caused by the different protocols used by software applications to communicate over the Internet, which give rise to the different form and functions apparent to the end user. This kind of protocol separation is intuitively apparent from December's map.

Since the map was drawn, the nature of the Internet has changed markedly, with certain information spaces dying off as they fall out of favor with users, particularly WAIS and Gopher. Undoubtedly, the biggest change has been the inexorable and exponential growth of Web space, which would now be a huge blue "blob" on the map, squeezing out and submerging many other information spaces. Other important information spaces have evolved and grown in prominence, such as instant messaging (e.g. ICQ), chat environments (e.g. IRC), multi-user game spaces (e.g. Quake) and streaming media (e.g. RealNetworks, MP3s). Also, large intranets have proliferated, creating important private information spaces.

December admits that: "In a way, my 1994 map was very naive – I didn't really have a clear idea of global nets at the time, but I intuitively pictured the conceptual relationships much like ancient tribes did – I knew that there was 'this place' and 'another place' and ways to 'get' from one place to the other."

Despite the naiveté of the map, it still provides one of the best conceptual views of cyberspace and today it has added historical interest.

3.1: Conceptual map of Internet information spaces
chief cartographer: John December (December Communications, Inc., Milwaukee, Wisconsin).
aim: to show the multiple, linked information space of the Internet, circa 1994.
form: Venn-diagram style, with different spaces shown as irregularly shaped blobs, key landmarks marked with symbols, and connections between spaces shown by red arrows.
technique: a hand-crafted diagram.
date: December 1994.
further information: see <http://www.december.com>

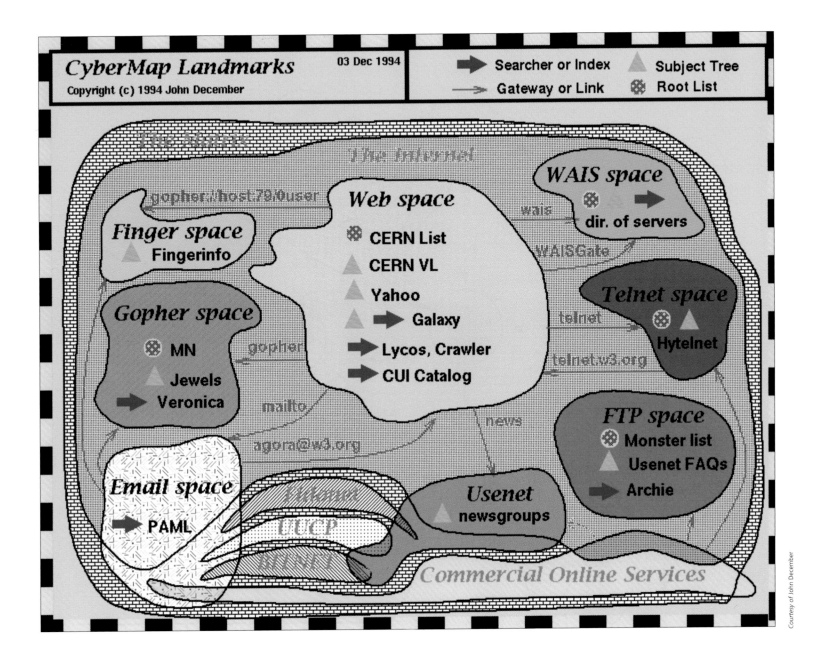

CyberMap Landmarks
03 Dec 1994
Copyright (c) 1994 John December

Searcher or Index
Gateway or Link
Subject Tree
Root List

The Matrix
The Internet

gopher://host:79/0user

Finger space
▲ Fingerinfo

Web space
❋ CERN List
▲ CERN VL
▲ Yahoo
▲ ➤ Galaxy
➤ Lycos, Crawler
➤ CUI Catalog

WAIS space
❋ ➤
dir. of servers

wais

WAISGate

Gopher space
❋ MN
▲ Jewels
➤ Veronica

gopher

Telnet space
❋ ▲
Hytelnet

telnet

telnet.w3.org

mailto

news

agora@w3.org

Email space
➤ PAML

Internet
UUCP
BITNET

Usenet
▲ newsgroups

FTP space
❋ Monster list
▲ Usenet FAQs
➤ Archie

Commercial Online Services

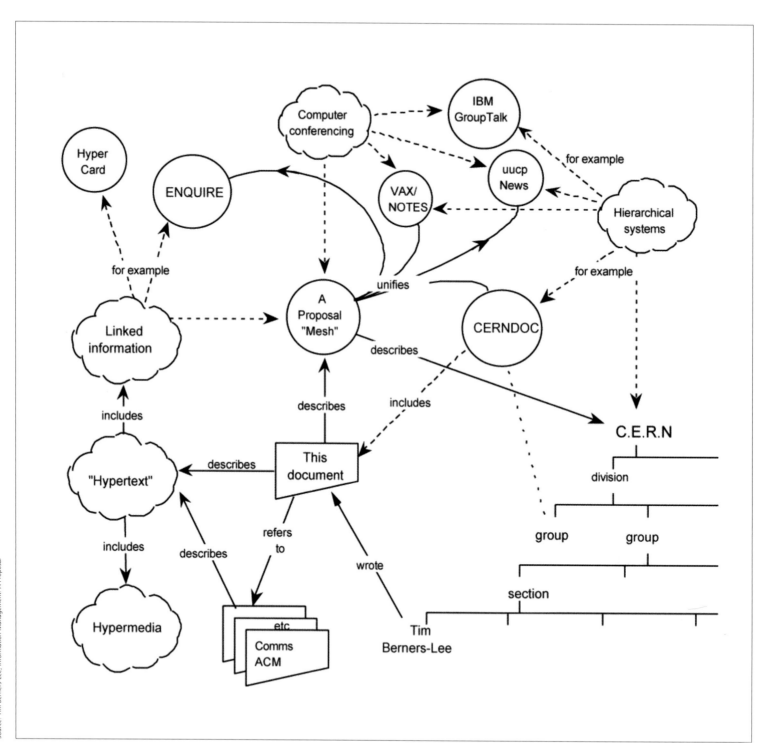

The beginning of the Web

It seems appropriate to continue by examining the conceptual map used by Tim Berners-Lee in his original proposal for a new hypertext system that would become the World Wide Web. Such spatializations are commonplace in the development of new systems, allowing designers to document the complex links between system components.

Berners-Lee invented the Web in 1990 while working at CERN, the European particle physics laboratory in Switzerland. The conceptual map was an important part of a document entitled *Information Management: A Proposal* written in 1989 to persuade CERN management to support the project to design and build a global hypertext system to better manage and maintain the large, complex and ever-growing information resources of that organization. The diagram opposite is a simple "mental map" of the ideas and concepts of the project using basic "circle and arrows" notation. The hypertext system at the heart of the diagram was called "Mesh" and it was only later that its name was changed to the World Wide Web once software development commenced in 1990. The idea was to provide a practicable and useful system that could seamlessly unify existing information resources and systems, and also be an open system to allow future development and growth. The ideas were built on many years of research into hypertext and information systems.

From the humble beginnings of this diagram and the supporting proposal, a rudimentary, but workable, text-only hypertext system emerged in the early 1990s. By the start of 1993, there were only perhaps 50 or 60 Web servers in the world. A critical turning point came in November 1993, when the Mosaic browser, a free piece of software developed by the National Center for Supercomputing Applications (NCSA), was released, providing a graphical way to navigate the Internet. Since Mosaic's release, the Web has grown tremendously to become a huge information space, offering all manner of information and services, and spawning multi-billion-dollar industries. Subsequently, commercial browsers such as Netscape and Internet Explorer have become the main means by which people surf the Net, but those browsers are still based fundamentally on the open, distributed and scaleable hypertext system envisioned by Berners-Lee whereby anyone was free to contribute a Web page or two and link to other resources. Utilising the power of hypertext, the Web rapidly became one of the most significant information and communication developments of the twentieth century.

3.2: The original conceptual design of the World Wide Web

chief cartographer: Tim Berners-Lee (while at CERN, Switzerland).
aim: to show the potential of a global hypertext system ("Mesh", later christened the World Wide Web) to unify disparate information resources in a large and dynamic organization.
form: a simple black-and-white "circles and arrows" diagram, sketching interrelations between ideas and concepts.
technique: a hand-crafted diagram.
date: March 1989.
further information: *Information Management: A Proposal*, by Tim Berners-Lee, at <http://www.w3.org/History/1989/ proposal.html>
further reading: *Weaving the Web: The Original Design and Ultimate Destiny of the World Wide Web by Its Inventor*, by Tim Berners-Lee (HarperBusiness, 1999).

Mapping individual websites

As soon as the Mosaic browser was released, the Web began to grow exponentially, with tens of thousands and then millions of new pages being added annually. Almost immediately, one of the key problems was trying to navigate efficiently and to search for particular pieces of information both within and across websites. Not unsurprisingly in both cases, a key strategy has been to explore and adopt spatializations as navigation tools, alongside keyword search tools and navigation bars. In this and the next section, we examine spatializations of individual websites, first detailing site maps and then mapping tools used in site planning, development and maintenance.

"Where am I?" is a question asked by most people when landed in the middle of a large website by a search engine, often followed by, "What else is available on this site?" and, "How do I get to that other information?" The site map is one of the key tools that site designers can provide to help surfers answer these questions and successfully navigate through their site. At their simplest, site maps are like a book's table of contents, while more sophisticated examples use advanced, interactive spatializations. In all cases, site maps aim to communicate, on a single screen, a site's content and enable a user to reach it with a single click. They are becoming increasingly important as websites become larger and more complex and are nowadays a common element of Web design, especially for large corporate sites. This is not to say that all of them are successful in their task or aesthetically pleasing to view.

The art and science of creating intuitive and useful website maps is still in its infancy. One of the most knowledgeable people in the nascent field of website mapping is Paul Kahn, a founding partner in the information design firm named Dynamic Diagrams. He commented: "I think website mapping is bouncing back and forth between two poles: it is absolutely necessary and it is impossible." At present, too many site maps fail in their attempts to guide disorientated surfers to their destination. This failure is due to a number of difficult problems, such as deciding on the most appropriate level of detail for effective communication, trying to balance local detail necessary for practical navigation, whilst also providing a global overview of the entire site.

Moreover, there are many possible elements that could be used to describe a site, all of which could be spatialized: page title, URL, screenshots, depth from home, or access restrictions. Further, the site maps themselves need to load quickly and conform to standard screen resolutions. Then there are the problems of keeping maps up to date on a dynamic site.

The plate opposite is a simple "table of contents", functional site map of Apple Computer's website as at August 2000. It employs no spatialization, consisting purely of simple text hyperlinks to key site content, where the spatial layout of the elements has no inherent meaning. The blue hyperlinks are grouped into 11 major sections, which are laid out on the screen in three rows, the aim being visual clarity. It is a simple and effective way of showing, on a single fast-loading screen, a high-level summary of all the content of Apple's huge website. This "table of contents" style is by far the most common form of site map employed on the Web today. This is because it is an effective mode of communication, but more crucially it is easily created and maintained.

3.3: A typical "table of contents" style of website map from Apple
chief designer: unknown (Apple Computer, Inc.).
aim: to provide a simple overview listing the major content areas of Apple's extensive website.
form: clean-looking hierarchical table of contents, with 11 major groups in three rows.
technique: simple clickable hypertext links take a user to an appropriate section of a site.
date: screenshot taken in August 2000.
further information: try the current version at <http://www.apple.com>
further reading: *Mapping Websites: Designing Digital Media*, Paul Kahn (Rotovision, 2000).

Site Map

Worldwide Apple Sites ▾ Go

About Apple

- Media & Analyst Info
- Job Opportunities
- Investor Relations
- Contact Information
- Apple & the Environment
- Privacy Policy
- Legal Information
- Web Badges

News & Events

- Hot News
- Apple eNews
- Features
- Seminars
- User Groups
- AppleMasters

QuickTime

- Movie Trailers
- QuickTime TV
- Features
- QuickTime Streaming Server
- Download QuickTime

iTools

- iTools
- KidSafe
- iDisk
- iReview
- iCards

Market Areas

- Education
- Creative
- Small Business
- Games

The Apple Store

- The Apple Store
- Apple Store Worldwide
- Apple Store for Education

U.S. Retail & Service

- Find a Dealer
- Loans & Leasing
- Product Registration

Support

- Product Support
- Protection Plan
- Technician Training
- Software Updates
- Product Specifications
- Discussions
- Tech Info Library
- Apple Solution Experts

Hardware

- iMac
- iBook
- Power Mac G4
- Power Mac G4 Cube
- Macintosh Server G4
- PowerBook
- Displays
- Pro Keyboard
- Pro Mouse
- AirPort
- Firewire

Software

- AppleShare IP
- AppleWorks
- AppleScript
- ColorSync
- Final Cut Pro
- iMovie
- Mac OS
- Mac OS X Server
- QuickTime
- Sherlock
- WebObjects

Developer

- Apple Developer Connection
- ADC Membership
- Technical Information
- Business Opportunities
- Apple Solution Experts
- Macintosh Products Guide

Courtesy of Apple Computer, Inc.

More interesting than simple text maps are the use of spatialized site maps. The following site maps use a metaphorical approach to represent the content of a website, employing a familiar image to aid navigation. The first example, Yell Guide's website (part of the UK's Yellow Pages website), used the visual styling of the London Underground map to create a site map. Active in 1997, the spatialization used four color-coded lines to represent different sections of the site (e.g. red for London and green for gardening), and the standard London Underground map symbol for an interchange station to represent individual pages.

Although interesting visually, its effectiveness as a navigation aid when compared with a conventional "table of contents" listing is debatable. Moreover, the spatialization is not a scaleable solution, having to be redrawn to accommodate growth. As a consequence, it was soon dropped from the site and has been replaced by other navigation tools.

3.4: The Tube map as a metaphorical site map
chief cartographer: not known.
aim: to provide an attractive overview map on the Yell Guide website (British Telecommunications plc.) showing major content areas on a single screen.
form: borrows the strong styling of the famous London Underground map.
technique: simple clickable hypertext links take a user to that section of the site.
date: 1997.
further information: this site map is no longer online.

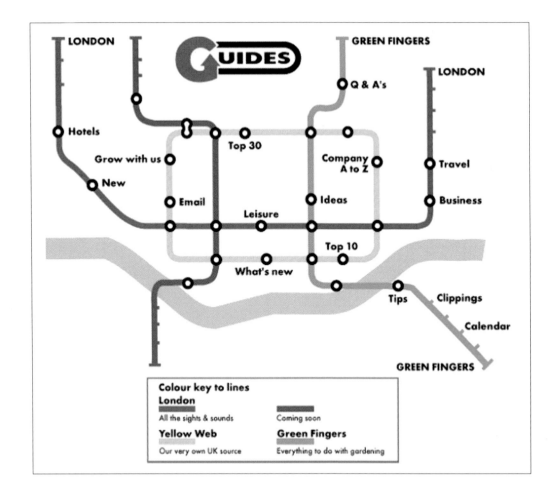

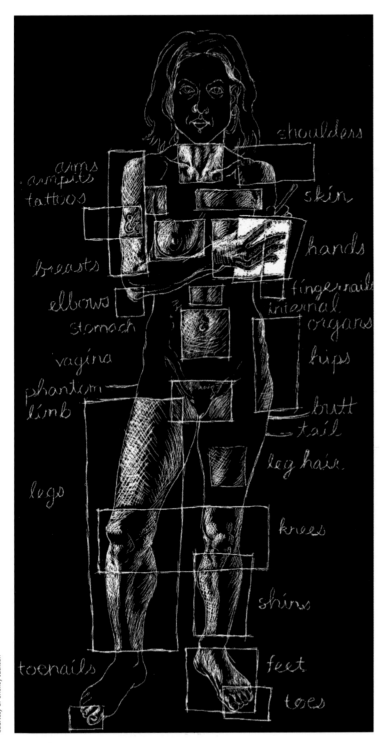

Another metaphorical site map is this striking and simple black-and-white chalk-style drawing of the human body, which is used to represent the content of a site. The "my body" site map is from Shelley Jackson's Web artwork entitled *My body, a wunderkammer* (a wunderkammer being a cabinet of curiosities, the forerunner of the modern museum, albeit with an emphasis on typology rather than chronology). It is more than just a functional map; it is an integral part of the actual artistic content of the website.

Shelley Jackson is an artist and writer who has experimented with hypertext writing and her "my body" site provides a hypertext narration on her feelings, memories and experiences of her body. The evocative sketch of the body used for the site-map is a self-portrait. Parts of the anatomy on the map, which are boxed and labeled, are linked to individual pages. Clicking on the "shins" area of the body map takes one to the page concerning her shins, described as follows: "My shins are slightly bumpy and when I get a tan, mottled with tiny white dots as with sunlight through leaves. The dots are faded scars, dozens of them of different vintages, criss-crossed and overlaid…". Here, the map is the territory it seeks to represent.

3.5: "my body" as a metaphorical site map

chief cartographer: Shelley Jackson.
aim: to provide an overview map showing major content areas on a single screen, but also to act as an integral part of the actual content of the site.
form: stylized black-and-white sketch of a female body.
technique: simple clickable hypertext links take a user to a selected section of the site.
date: 1997.
further information: "my body" at <http://www.altx.com/thebody/>
further reading: Shelley Jackson's homepage, entitled "Ineradicable Stain", at <http://www.ineradicablestain.com>

This example, Site Lens, similarly aims to provide a site map designed to help visitors find particular Web pages. Unlike the previous site maps, however, it uses powerful interactive graphics to show the extent of available pages and the overall hierarchical structure of a site. It has been developed by Inxight Software, a spin-off from the famed Xerox Palo Alto Research Center (whose work we will look at more closely later in this chapter). Site Lens is based on information visualization research in the mid-1990s that developed fish-eye distortion techniques in order to produce a hyperbolic browser of hierarchical graphical structures.

The technique works by warping the spatial view of the data under the "lens", so that elements at the center of the map appear much larger than those at the periphery. The user is able to grab page objects and drag them to the center of the map, where that area will be enlarged smoothly with an animated transition. In this manner, a user can explore a large hierarchical website of hundreds of pages with simple point-and-click browsing, with local detail in the center of the map and wider context visible in the surrounding area. Indeed, the underlying technique is often described as "focus + context". Individual pages are usually represented in the Site Lens by a rectangular block containing the title (but it could include an illustrative icon or thumbnail image). Double-clicking on a title or icon will load the corresponding page into the main browser window.

The figures displayed below and opposite show different examples of Site Lens maps, illustrating various aspects of this mapping system. The main images displayed show a stylistic demo of the Porsche website. The large image opposite is the default view and is centered on the Porsche homepage. Organized in a circle around this home page information is the branching structure of different segments of the site. In the second screenshot, a user has chosen to look in more detail at the "Boxster" portion of the site (one of Porsche's models), and so the user has selected and dragged that node into the center. When it is at the heart of the lens, its subsidiary pages grow and become legible. Moving further down the "Boxster" path, the third screenshot focusses on the "Engine" page.

The two other examples are Site Lens demos of the online toystore eToys.com, which uses many thumbnail images of toys to illustrate the different sections of its site, and Lexis-Nexis, one of the world's largest information archives. In the latter case, the site map provides a way to browse the 22,000 information sources that the corporation has access to – although, in this example, the map has become overly dense at certain branches of the hierarchy and is reaching the limits of its capability to present a clear visualization of the information.

3.6: Site Lens interactive fish-eye site maps

chief cartographers: Inxight Software, based on research by John Lamping, Ramana Rao and Peter Pirolli (User Interface Research Group, Xerox Palo Alto Research Center).
aim: to provide an interactive overview map showing major content areas of the website on a single screen.
form: arc–node tree structure viewed through a simulated fish-eye lens.
technique: hyperbolic-tree visualization.
date: initial hyperbolic browser developed in 1995, and commercial product Site Lens Studio launched in August 1999. Screenshots taken in August 2000.
further information: <http://www.inxight.com>
further reading: "A focus + context technique based on hyperbolic geometry for visualizing large hierarchies", by John Lamping, Ramana Rao, and Peter Pirolli. ACM Conference on Human Factors in Software (CHI '95), Denver, Colorado. <http://www.acm.org/sigchi/chi95/proceedings/papers/ jl_bdy.htm>

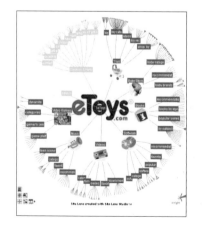

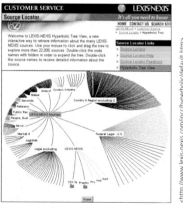

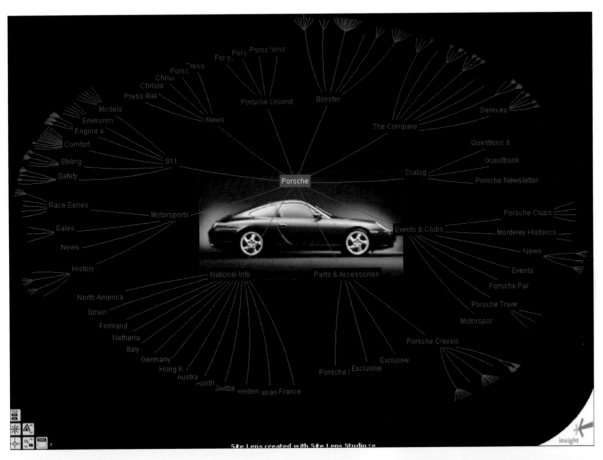

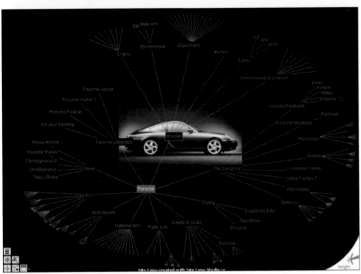

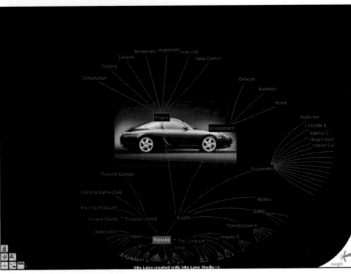

The second interactive site mapping system we detail is SiteBrain, developed by TheBrain Technologies Corporation. It too uses a hierarchical graphical approach to interactively map the structure of a site. However, it does not use fish-eye distortion and therefore cannot show the whole structure of a site on a single page. Instead, it relies on interactive graphics to enable users to navigate hierarchical pathways. The example screenshots show the application of SiteBrain to mapping a large online magazine called *Mappa.Mundi*.

The first screenshot top-left shows the top of the website's hierarchical structure with the homepage highlighted. Sibling nodes show the organization of the four main categories of content on the site. These represent four different ways to navigate to actual magazine articles. For example, you can access the articles based on who wrote them (authors) or by date (monthly edition of the magazine) or by sections or by topics. Clicking on the "Sections" topic causes the map to be smoothly redrawn, bringing this topic to the center of the

spatialization and revealing the various sections of the magazine as "children" positioned below the "sections" heading, whose siblings are positioned to the right-hand side (top-right). Exploring further into the map by clicking on the "map of the month" section leads to another transformation to reveal a further level in the hierarchy (bottom-left). The individual articles of the "map of the month" section are now listed as children nodes, while the right-hand sibling nodes are the other sections. Selecting the article, "A Map of Yahoo!" reveals that there are no further child pages in this hierarchy and, instead, two higher levels are shown: "February 2000" (bottom-middle) and "Map of the Month", with their siblings arrayed on the right-hand side. This offers a quick way to move sideways and access related information – for example, clicking the "February 2000" link redraws the map showing other articles that were published in that month (bottom-right). At any point, clicking on a particular item opens the appropriate Web page.

3.7: SiteBrain sitemap for Mappa.Mundi Magazine

chief cartographers: Corinne Becknell, Rebecca Hargrave, Marty Lucas, Carl Malamud at Mappa.Mundi using the SiteBrain system.
aim: to provide simple overview map showing major content areas on a single screen.
form: hierarchical graph structures.
technique: based on "concept mapping" technology of the brain.
date: SiteBrain was launched in 1999. Screenshots of *Mappa.Mundi* Magazine's SiteBrain output were taken in August 2000.
further information: TheBrain Technologies Corp. at <http://www.thebrain.com> Mappa.Mundi Magazine at <http://mappa.mundi.net>

Dynamic Diagrams has developed an innovative visual metaphor to represent website structure, which it calls the Z-Diagram. It is a 2.5-D landscape view, with Web pages represented as small standing cards laid out across an isometric plain. This provides a sense of depth without a vanishing point, so that objects are of a uniform size throughout the map. The metaphor was conceived by Krzysztof Lenk, a designer and founding partner in the company, to display the structure of a multimedia encyclopaedia. Detailed, handcrafted, Z-Diagrams are widely employed by Dynamic Diagrams as planning maps when designing and redesigning large websites.

Dynamic Diagrams has also used its Z-Diagram metaphor in an interactive site map utility for surfers. The system is called MAPA. Using MAPA, surfers can request a simple Z-Diagram style of map from any page on a website. The Z-Diagram will then be displayed in a small pop-up window displaying the localised structure. An example of a sequence of MAPA maps is shown opposite. Importantly, the MAPA system also provides "back-end" tools to analyze the structure of a large site and to derive the key hierarchical structure, which is stored in a database. MAPA is, therefore, a data-driven website map rather than a handcrafted one. The top image shows a screenshot of the MAPA map of Dynamic Diagram's website, with cards projecting vertically from the ground plain representing the pages. The cards are spatially arranged to reflect the dominant hierarchical structure of the site without cluttering the screen with multiple hyperlinks between pages. Differently colored carpets delineate different levels in the hierarchy.

Users can interact with the MAPA map in several ways. For example, passing the cursor over a card will cause it to be highlighted and for the title to be displayed as a flag. Users can also navigate within the map itself by refocussing the layout around a different page. This is achieved by clicking on cards with a dark bar on them, whereupon the pages rearrange themselves in a smooth animated transition – the bottom plates provide snapshots of this. This is one of the key "wow factors" of MAPA, according to Paul Kahn, who comments: "Moving from one view to the next without the animation would be incomprehensible. With the animation, most people can get it. And it is fun to watch the pages march across the screen and rise up out of the carpet." Finally, MAPA can be used to move the user's browser to a different page in the site simply by double-clicking on a card of interest. The MAPA system provides one of the most innovative and effective interactive website maps currently available.

3.8: MAPA Z-factor site map of Dynamic Diagrams

chief cartographers: David Durand and Paul Kahn (Dynamic Diagrams, USA).
aim: to provide interactive site map to aid navigation.
form: a Z-factor animated map with pages represented by small cards. These are arranged in a hierarchical structure by position and color-coding.
technique: custom-written applications to analyze site structure and create a suitable hierarchical representation. Java applet used in order to deliver an interactive map, centered on current page.
date: initially developed in 1997. Screenshots taken in August 2000.
further information: MAPA at the Dynamic Diagrams site at <http://www.dynamicdiagrams.com>
further reading: *Mapping Websites: Designing Digital Media* by Paul Kahn (Rotovision, 2000). Technical paper "*MAPA: A system for inducing and visualizing hierarchy in websites*" by David Durand and Paul Kahn, Ninth ACM Conference on Hypertext and Hypermedia, June 1998; <http://www.dynamicdiagrams.com/pdf/papers/mapaht98.pdf>

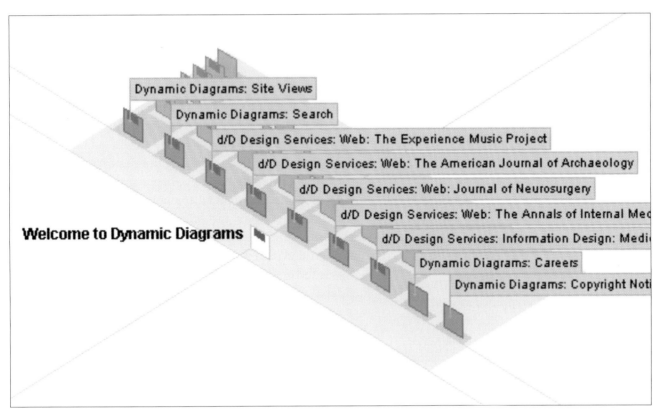

Dynamic Diagrams: Site Views

Dynamic Diagrams: Search

d/D Design Services: Web: The Experience Music Project

d/D Design Services: Web: The American Journal of Archaeology

d/D Design Services: Web: Journal of Neurosurgery

d/D Design Services: Web: The Annals of Internal Med

d/D Design Services: Information Design: Medi

Dynamic Diagrams: Careers

Dynamic Diagrams: Copyright Noti

Welcome to Dynamic Diagrams

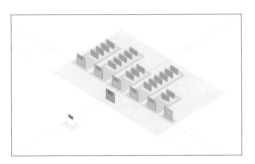

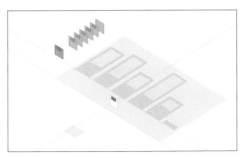

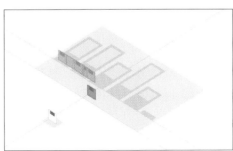

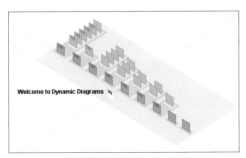

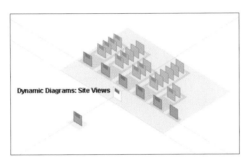

Mapping tools to manage websites

Specialized website-mapping applications that are aimed at those who manage large sites are generally analysis and management tools rather than navigational aids. Their role is to reveal the underlying structure of websites by detailing the content and links between pages. In this section, we detail a number of examples, starting with hand-crafted diagrams before moving on to discuss interactive tools that provide detailed and sophisticated interactive spatializations.

Dynamic Diagrams specializes in consulting on information architecture for the design or redesign of large websites. A key tool it uses in working with its clients is high-quality planning maps showing proposed structures and pathways through Web pages. These are hand-crafted, taking skilled designers many hours to put together, and they are often printed as large poster maps to use as discussion tools in meetings. The most distinctive mapping style developed by Dynamic Diagrams is undoubtedly its Z-form diagrams. These use the same spatial concepts and representation as the company's interactive MAPA package.

A good example of a Z-form diagram is that of the *Nature* overview diagram produced in July 1999 (opposite). It displays in a clear and concise manner the key elements of *Nature's* information architecture on a single page, showing the key Web pages, routes, click depth and access restrictions. The site begins with the homepage at the bottom left and then pages at one, two and three clicks' depth are drawn progressively further back toward the top right-hand corner of the page. On this particular website, not all content is free to access, and so, in order to show clearly the different access rights across the site, both height and strong color-coding are used. The green area at ground level is free for all visitors. The next level of access is the orange raised area, which is restricted to registered users. Full access to the journal articles is restricted to paying subscribers, and this is represented by the red plinth. Finally, there is some special language content that is limited to Japanese subscribers. Another useful feature of the map is the linking of thumbnail screenshots of certain key Web pages to their position in the structure.

3.9: Z-form website planning diagram

chief designers: Chihiro Hosoe, Nancy Birkholzer, Paul Kahn and Krzysztof Lenk (Dynamic Diagrams, USA).
aim: to aid the planning and design of the structure of the *Nature* website.
form: Z-form representation where Web pages are standing cards on an isometric plane. The distance from bottom-left to top-right shows the click depth, and the height above the ground plane represents increasing access permissions.
technique: hand-crafted diagram.
date: July 1999.
further information: <http://www.dynamicdiagrams.com>
further reading: *Mapping Websites: Designing Digital Media* by Paul Kahn (Rotovision, 2000).

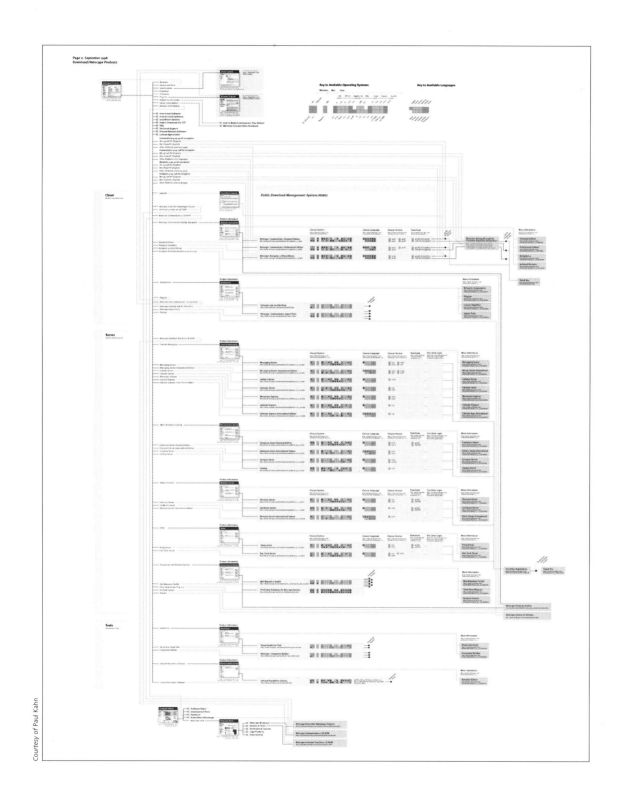

Another example of Dynamic Diagrams' large-scale website maps for planning and design is the Netscape software download-area diagram. This was produced as part of a consulting project for Netscape, analyzing how its site was utilized by browsers. The style of mapping used is very different from the Z-form and is reminiscent of a wiring diagram for an electrical circuit-board. The aim of the map was to highlight the route users needed to take to download different pieces of software – quite a complex task given all the different permutations of operating system, language and encryption levels that Netscape offers.

The start point for the map and the website is the download homepage, shown at the top-right. The map is divided vertically into sections based on three broad categories of software: client (browsers), server and tools. A user then follows a given pathway running from left to right on the map, selecting an operating system and language before finally reaching the "download point" from where the software is transferred. These points are indicated on the map by small blue dots. The designers of the map have developed a compact but effective way of representing the complex selection process as a series of check-boxes.

3.10: Wiring diagram of a website

chief designers: Piotr Kaczmarek, Paul Kahn, and Krzysztof Lenk (Dynamic Diagrams, USA).
aim: to show the possible routes to download a multitude of different software available from the Netscape website.
form: detailed wiring diagram.
technique: hand-crafted diagram.
date: September 1998.
further information: see <http://www.dynamicdiagrams.com>
further reading: *Mapping Websites: Designing Digital Media* by Paul Kahn (Rotovision, 2000).

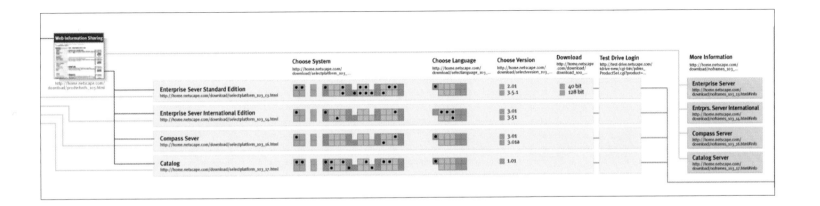

The final Dynamic Diagrams example that we examine is a detailed strategic-planning map that summarizes the relationship across websites owned by a single corporation – in this case the large international publishing corporation named Verlagsgruppe Georg von Holtzbrinck. This corporation owns an array of public websites and the spatialization shown here summarizes a great deal of information about the sites: who operates them, what content they hold, and how they are interlinked.

Each website is represented by a colored circle with various icons summarizing key attributes of the site. Websites are color-coded and spatially arranged in the map according to the publishing division that operates them. Each circle contains a thumbnail image of a screenshot of the website's homepage, along with small icons indicating the company type (newspaper, scientific publisher, television station, etc.), the country of operation (shown by a small flag), and the language used at the site. Particular content and features (such as online advertising, catalog searching) are indicated by further icons on the bottom half of the circles. The spatial arrangement of the circles is reminiscent of a chemistry periodic table. Navigational links between websites are shown by the fine lattice of connecting lines. The color of the lines, their thickness, and the kind of arrow at their termini all carry information as well. A one-way hyperlink is represented by a thin line, whilst a bilateral connection is shown by a thicker line. Lastly, an alphabetic catalog of names at the top of the map is coded to a labeled grid to help the reader locate individual websites.

3.11: Strategic map of the multiple websites of the Holtzbrinck Corporation

chief designers: Piotr Kaczmarek, Paul Kahn, Krzysztof Lenk, Magdalena Kasman, Nancy Birkhölzer and Angelika Binding (Dynamic Diagrams, USA).
aim: to produce a strategic map showing 130 public websites run by major publishing corporation Verlagsgruppe Georg von Holtzbrinck.
form: complex wiring diagram.
technique: large wall poster map, hand-crafted.
date: May 1999.
further information: see <http://www.dynamicdiagrams.com>
further reading: *Mapping Websites: Designing Digital Media* by Paul Kahn (Rotovision, 2000).

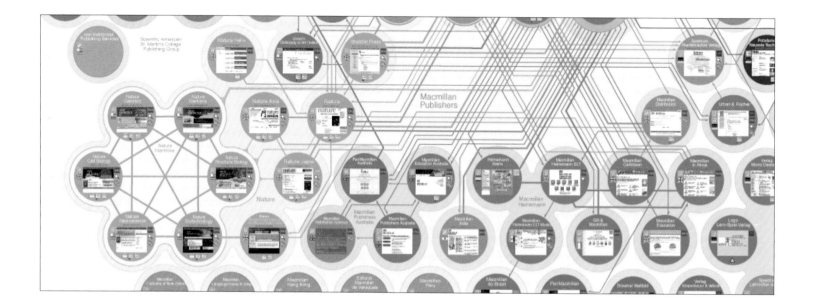

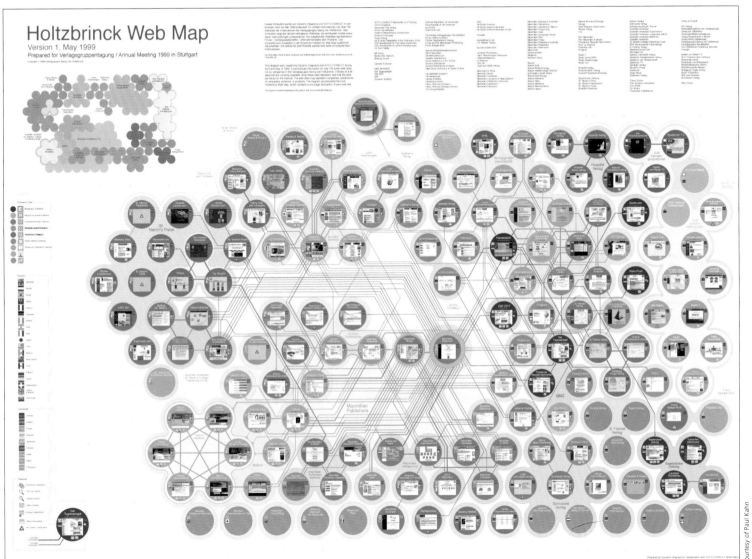

Holtzbrinck Web Map

Version 1, May 1999

Prepared for Verlagsgruppentagung / Annual Meeting 1999 in Stuttgart

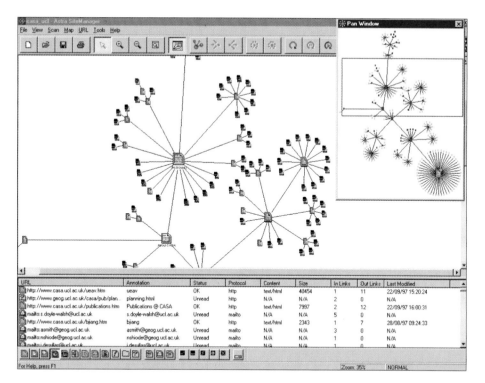

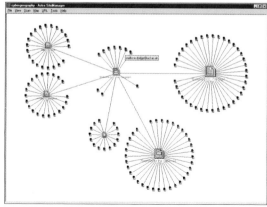

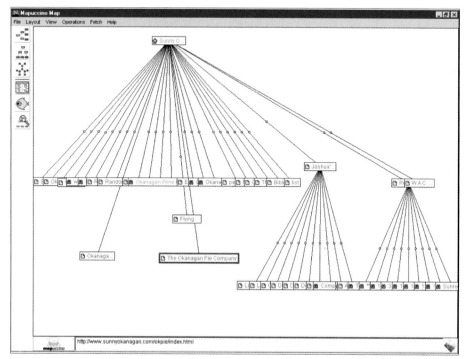

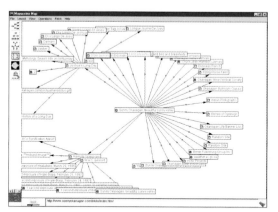

Websites with rapidly changing content can be difficult to manage. For example, preventing broken links, managing page changes, tracking Web page sizes to ensure a reasonable download time, and understanding the interconnections of a website can be time-consuming tasks. One method to try to address these problems is to employ an interactive, visual website-management tool. In the following examples, we detail three such tools.

The top-left and top-right images opposite display the interface of Astra SiteManager. SiteManager is a standalone application that allows a manager to pinpoint website problems (such as broken links), to run Web scans, and to improve the organization of a site by revealing hotspots (achieved by monitoring traffic through a site).

The screenshot displayed top-left is a mapping of the Center for Advanced Spatial Analysis's website. The pages are represented by nodes and the links between pages by lines. In the top right-hand corner is a small overview map that can be panned to see sites too big to fit onto a single screen. At the bottom of the screen, details about each node in the graphic are listed.

The image shown below displays the interface of FrontPage 98 Explorer. It is a companion utility to Microsoft's popular FrontPage HTML (Hypertext Markup Language) authoring package, and it allows users to examine the hierarchy of their pages. The interface is divided into two, with the left-hand frame displaying link hierarchy and the right-hand one a graphical representation of the links, which can be expanded or compressed. If a page's contents are altered, the package can be run to determine whether links between pages have been severed.

The final example, displayed bottom, opposite, is Mapuccino. This was developed by researchers at IBM's Haifa Research Lab in Israel. It allows users to capture and view the overall structure of any online website, and to navigate visually through the site's contents. It provides five mapping modes (tree vertically, tree horizontally, tree as a star, list, and fish-eye view). The example screenshots show a simple vertical tree revealing the site's hierarchy of pages (bottom-left) and a fish-eye view of the same site (bottom-right).

3.12: Astra SiteManager

chief cartographer: Mercury Interactive Corporation.
aim: to provide a suite of visual Web management tools.
form: arc–node, hierarchical circular tree structures.
technique: displays page information and the links between pages as a scalable graph.
date: 1998.
further information: see <http://www.merc-int.com/products/astrasitemanager/>

3.13: FrontPage 98 hyperlink view

chief cartographer: Microsoft.
aim: to provide a visual Web management tool as part of a larger HTML editing application.
form: a hyperlink view, centered on the selected page, showing incoming and outgoing hyperlinks.
technique: a horizontal tree graph.
date: 1998.
further information: see <http://www.microsoft.com/frontpage>

3.14: Mapuccino

chief cartographer: IBM Haifa Research Lab.
aim: to provide a visual Web management tool.
form: provides several different graph layouts of the pages and hyperlinks, including tree structures and fish-eyes.
technique: custom-written Java application.
date: 1997.
further information: see <http://www.ibm.com/java/mapuccino/>

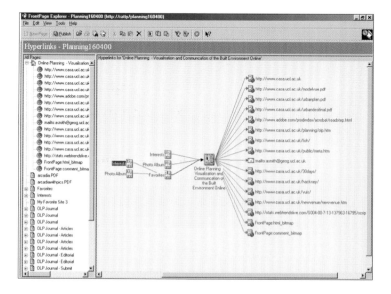

The visual Web tools so far discussed are generally designed to help manage small-to-medium-sized sites. NicheWorks is a tool created for exploring and managing very large sites of 10,000+ nodes and, unlike the previous packages, also has many other applications (e.g. telephony and telephone fraud, software analysis, email patterns, information retrieval, medical data and market-based analysis). Originally a standalone tool, it is now part of the EDV (Exploratory Data Visualizer) environment.

In order to determine the structure and status of nodes (pages), a webcrawler is used. This information is then applied in order to construct large graphs using sophisticated spacing routines. Three layouts are possible: circular, hexagonal and tree. The first graph (below) uses the hexagonal layout of a typical website. Here, orange squares represent local pages; orange

circles – local images; blue squares – offsite pages; yellow squares – CGI queries; light blue – avoided (generally email). In addition, key nodes have been labeled. The second graph (opposite) employs the circular layout and is a graph of the *Chicago Tribune* website.

3.15: NicheWorks mapping of a large website

chief cartographer: Graham Wills (Lucent Technologies – Bell Labs, USA).
aim: to explore the site structure of networks and sites with 10,000+ nodes.
form: grouped, color-coded arc–node graphs.
technique: uses MOMspider tool to determine links between nodes and edges between groups; custom software to construct visual graphs.
date: 1997.
further information: see <http://www.bell-labs.com/~gwills>
further reading: "NicheWorks – Interactive Visualization of Very Large Graphs", by Graham J. Wills, 1997, in *Graph Drawing "97 Conference Proceedings*, pp. 403–14, Rome, 1997, Springer–Verlag. <http://www.bell-labs.com/user/gwills/NICHEguide/nichepaper.html>

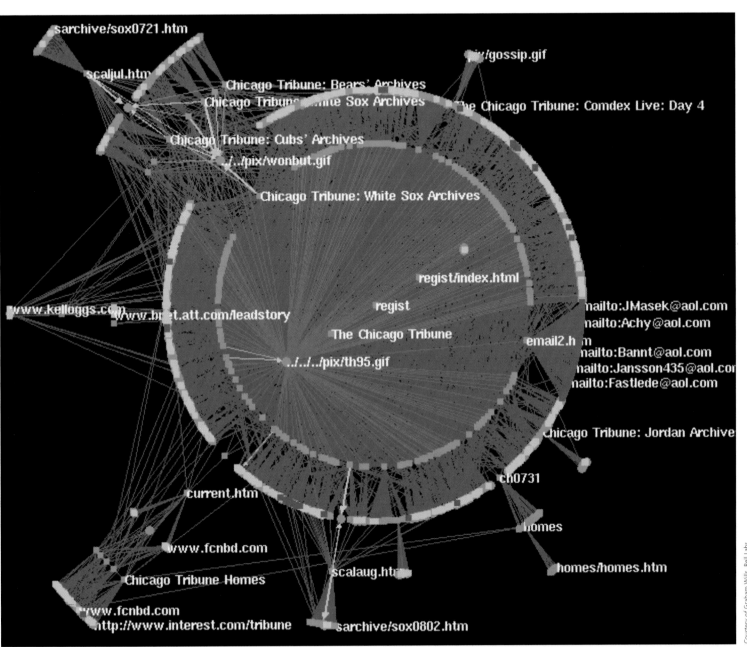

sarchive/sox0721.htm

pix/gossip.gif

scaljul.htm

Chicago Tribune: Bears' Archives

Chicago Tribune: White Sox Archives

The Chicago Tribune: Comdex Live: Day 4

Chicago Tribune: Cubs' Archives

.../.../pix/wonbut.gif

Chicago Tribune: White Sox Archives

regist/index.html

regist

www.kelloggs.com

www.bret.att.com/leadstory

mailto:JMasek@aol.com

mailto:Achy@aol.com

The Chicago Tribune

email2.htm

mailto:Bannt@aol.com

mailto:Jansson435@aol.com

mailto:Fastlede@aol.com

.../.../.../pix/th95.gif

Chicago Tribune: Jordan Archive

ch0731

current.htm

homes

www.fcnbd.com

homes/homes.htm

Chicago Tribune Homes

scalaug.htm

www.fcnbd.com

http://www.interest.com/tribune

sarchive/sox0802.htm

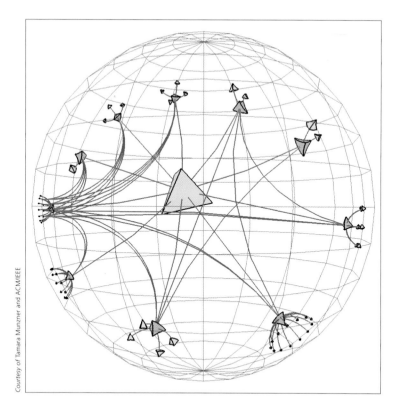

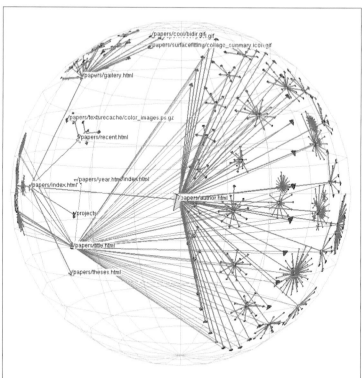

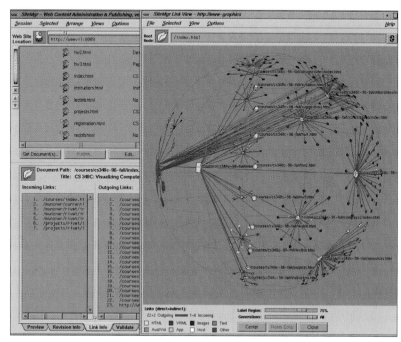

In contrast with Euclidean 3-D spaces, Tamara Munzner, then a graduate student in the Computer Graphics Laboratory, Stanford University, has investigated the potential of constructing spatializations in hyperbolic space. Hyperbolic spaces have advantages for visualizing the detailed structure of large graphs containing many thousands of nodes, such as the Web. As Munzner and research collaborator Paul Burchard comment: "The felicitous property that hyperbolic space has 'more room' than Euclidean space allows more information to be seen amid less clutter, and motion by hyperbolic isometries provide for mathematically elegant navigation." The figures opposite display three examples of Munzner's 3-D hyperbolic spaces.

The spatializations provide a novel way for exploratory visual browsing of the page–hyperlink structures of large websites. The structure of nodes and links is projected in hyperbolic space inside a ball, known as the "sphere at infinity" (top-left). The user is able to manipulate the graph, rotating and spinning it inside the sphere. Like the fish-eye distortion technique that we looked at in relation to Inxight's Site Lens spatialization, hyperbolic space gives greater visual presence (in terms of screen-space) to elements at the center of the view. As objects are moved to the periphery, they smoothly shrink in size. At the edge of the sphere, the nodes are very small, but the user can easily drag them into the center to enlarge them and see them in detail. In this manner, the hyperbolic spatialization can provide a view of the detailed graph structure, whilst still showing the overall context.

An example by Munzner and Burchard from 1995 is shown in the large spherical image (top-right) that spatializes the structure of two layers of their department's website. The pyramid glyphs represent pages, and the curving lines are the principal hierarchical hyperlinks. This spatialization was part of a Web mapping system called Webviz, which could gather the structure of a specified portion of the Web and then visualize it in a 3-D viewer called Geomview. Munzner undertook further refinements in the underlying hyperbolic spatializations, developing what is known as the H3 layout algorithm and a more powerful viewing system (H3Viewer) that enables interactive exploration of their graphs of 100,000 or more nodes.

The image top-left is a spatialization of part of the Stanford graphics group website drawn as a graph in 3-D hyperbolic space. The entire site has more than 20,000 nodes, and some 4,000 of them in the neighborhood of the paper's archive appear in this image. In addition to the main spanning tree, the image shows the non-tree outgoing links from an index of every paper by title. The tree is oriented so that ancestors of a node appear on the left and its descendants grow to the right. Lastly, an example of the H3Viewer incorporated into a product called SiteManager from SGI is shown in the lower map on the opposite page. SiteManager is a tool for webmasters that provides a fluid and scaleable view of a website structure.

Munzner has also done interesting work on the geographic visualization of the Internet in mapping MBone, which we discussed in chapter 2 (page 37).

3.16: Visualizing websites in 3-D hyperbolic space

chief cartographer: Tamara Munzner (while at the Computer Graphics Laboratory at Stanford University, USA).
aim: to allow users to interactively analyze the structure and usage of a website.
form: a fish-eye view of 3-D graphics, representing the pages and hyperlinks of a website and projected inside a transparent sphere.
technique: hyperbolic visualization.
date: examples from her research, 1995–2000.
further information: see <http://graphics.stanford.edu/~munzner/>
further reading: "Interactive Visualization of Large Graphs and Networks", Tamara Munzner, PhD thesis, Stanford University, June 2000.
<http://graphics.stanford.edu/papers/munzner_thesis/>

Mapping website evolution

Nearly all of the mappings we discuss in this book fail to present the dynamically changing ecology of users and structures within the Net. In this section, we present Xerox PARC's examples of attempts to visualize the evolution of a website over a period of time. This form of analysis, although under-researched, is of importance because it allows a temporal analysis that has great potential utility for Web designers, allowing them to plan future site development to meet expected demand and to also manage their resources. It is also of use to network providers, allowing them to gauge where new potential network growth might occur.

The three figures shown here display the evolution of a single website over time. In the top image opposite, four disk-trees are shown in chronological order, each a discrete time slice collected over a four-week period in April 1997. A disk-tree is formed by analyzing a site's (in this case www.xerox.com) content (words in page and clusters of items), usage (frequency of page visits) and topology (hyperlink structure) (abbreviated to CUT), and plotting this data in a hierarchical circular pattern around its root page. The four disk-trees form a "time tube" presenting the site's evolution. They were created with Xerox's Web Ecology and Evolution Visualization (WEEV) software. By plotting the evolution over time along with usage, it is possible to investigate how well new material is being consumed, and to track and monitor particular trends. The frequency by which a page is accessed is denoted by line size and brightness. New pages are colored red, old pages are colored green, and deleted pages are yellow.

The cone-trees (bottom, opposite) also represent the evolution of xerox.com's website over the same four-week period. They were constructed using Xerox's Web Analysis Visualization Spreadsheet (WAVS). This software extends the CUT analysis to include a rating indicator that computes its similarity or relevance to other files. In the image, a rainbow color scale is used where red corresponds to pages with high usage and blue to those with low usage. The three rows correspond to a color threshold scale so that values between 0 and 100 are mapped in row 1, 100–500 are mapped in row 2, and 500–2,000 in row 3. The image shown below is a close-up view of part of the hierarchical structure.

3.17: Visualizing and analyzing website evolution

chief cartographers: Ed H. Chi, Stuart K. Card and colleagues (User Interface Research group at Xerox PARC, USA).
aim: to analyze the structure and usage of a website over time.
form: disk-tree displays arranged into a "time tube".
technique: analyzes a site's content, usability and topology at specific time slices and presents them in chronological order as a time tube.
date: 1997/1998.
further information: see <http://www.parc.xerox.com/istl/groups/uir/>
further reading: "Visualizing the evolution of web ecologies", Ed H. Chi, Jim Pitkow, Jock Mackinlay, Peter Pirolli, Ross Gossweiler, and Stuart K. Card, Conference on Human Factors in Computing Systems (CHI '98), Los Angeles, ACM: 400–407.
<http://www-users.cs.umn.edu/~echi/papers/chi98/1997-09-WWW-Vizualization4.pdf>
"Sensemaking of Evolving Web Sites using Visualization Spreadsheets", by Ed H. Chi and Stuart K. Card, in *Proceedings of the Symposium on Information Visualization* (InfoVis '99), pp. 18–25, 142, IEEE Press, San Francisco, CA, 1999.
<http://www-users.cs.umn.edu/~echi/papers/infovis99/ chi-wavs.pdf>

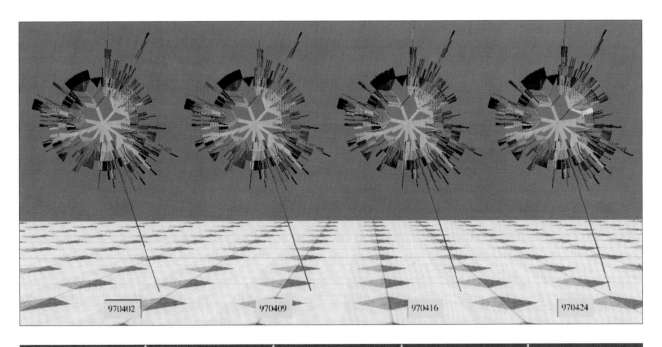

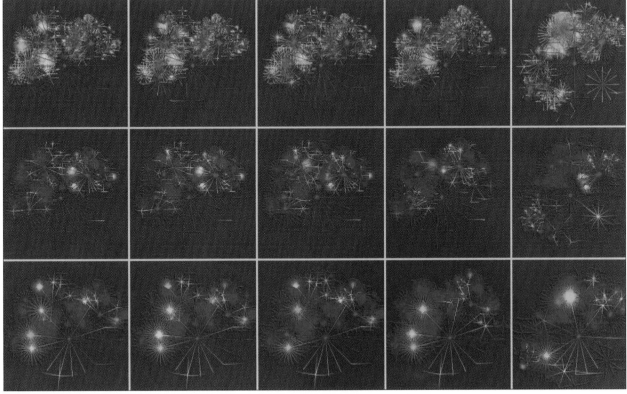

Mapping the web

103

Mapping paths and traffic through a website

In addition to mapping the extent and evolution of websites, another development has been the implementation of software to monitor specific user paths and traffic flows within and across sites. These data are important as they reveal how websites are being traversed by users and thus reveal the effectiveness (or otherwise) of site design and the popularity (or otherwise) of the material on it. A number of different utilities have been developed that analyze such data, as set out next.

3.18: VISVIP – mapping how users move through a website

chief cartographers: John Cugini and Jean Scholtz (National Institute of Standards and Technology, USA).
aim: to visualize the path of a user through a site.
form: 2.5-D representation, with paths displayed as spline curves.
technique: uses WebVIP to log the paths taken.
date: 1999.
further information:
see <http://www.itl.nist.gov/iaui/vvrg/cugini/webmet/visvip/vv-home.html>
further reading: "VISVIP: 3D visualization of paths through web sites", by John Cugini and Jean Scholtz, Proceedings of the International Workshop on Web-Based Information Visualization (WebVis 99), pp. 259–263, Florence, Italy, 1–3 September 1999.
<http://www.itl.nist.gov/iaui/vvrg/cugini/webmet/visvip/webvis-paper.html>

The VISVIP tool developed by John Cugini and Jean Scholtz visualizes the hierarchical structure of a website and the paths that users take through that site. Individual pages are denoted by boxes, and the connections between pages are shown as straight lines. Boxes are labeled with identifiers and color-coded by type: blue for HTML, purple for directories, green for images. The paths taken by users are shown as colored splines.

In the case of the main image (opposite), two paths are shown, one green and one red, each representing a single user. Dotted columns represent the time that each user spent on each page, with the taller the column meaning the more time spent browsing that page. Whilst the package has wider application, it has been developed to provide an experimental setting in which to test how users navigate and search through sites.

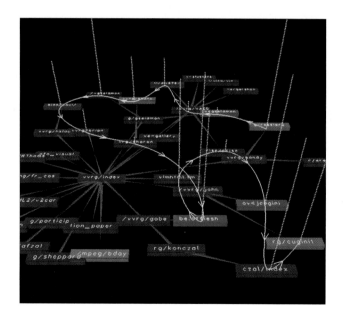

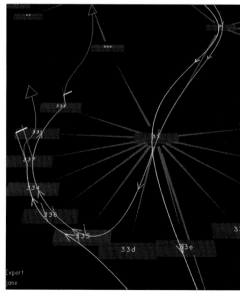

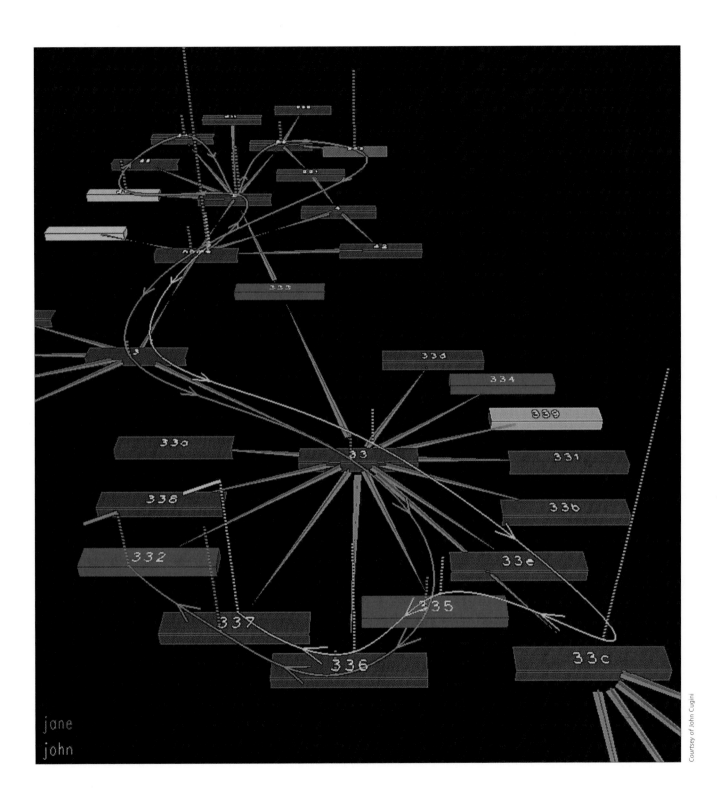

jane
john

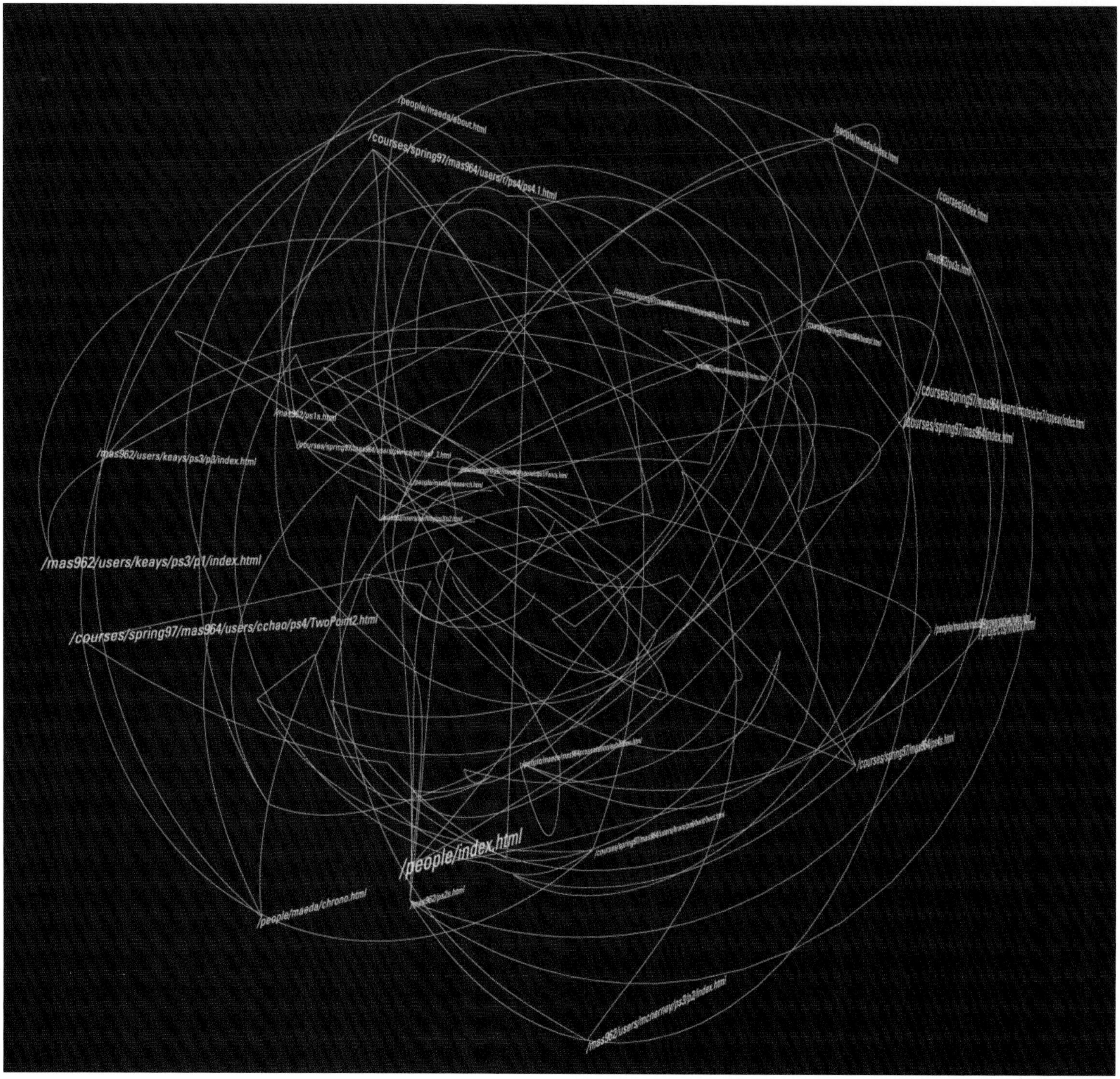

/people/maeda/about.html

/courses/spring97/mas964/users/r/ps4/ps4.1.html

/people/maeda/index.html

/courses/index.html

/mas962/ps3c.html

/courses/spring97/mas964/users/piperca/ps1/intro/index.html

/courses/spring97/ps4/about/index.html

/courses/spring97/mas964/users/ritzalay/ps7/appear/index.html

/mas962/ps1s.html

/courses/spring97/mas964/index.html

/mas962/users/keays/ps3/p3/index.html

/courses/spring97/mas964/users/piperca/ps1/ps1_1.html

/people/maeda/research.html

/mas962/users/keays/ps3/p1/index.html

/people/maeda/projects/index.html

/courses/spring97/mas964/users/cchao/ps4/TwoPoint2.html

/courses/spring97/mas964/users/francois/ps4/text/box.html

/courses/spring97/mas964/ps5.html

/people/index.html

/people/maeda/chrono.html

/mas962/ps3c.html

/mas962/users/mcnerney/ps3/p3/index.html

Organic Information Design, developed by Ben Fry through his package named Valence, is an attempt to produce 3-D dynamic representation of how a site is being used and traversed by browsers. As such, instead of a typical Web usage report composed of bar charts revealing such information as "20,000 people visited the home page", a self-evolving map of how people have been using the site is constructed. Fry details that this self-evolving map is driven by traffic patterns rather than the structure that the site's designer has put in place. The result is an ever-changing image of how the site's traffic evolves through time. Those sites that are linked are connected directly in the spatialization, with the sites most frequently visited placed on the outer parts of the composition.

3.19: Valence

chief cartographer: Ben Fry (Aesthetics & Computation Group, MIT Media Lab).
aim: to produce 3-D dynamic representations of how a site is being used and traversed.
form: 3-D, complex graphs.
technique: monitors site usage from the log and constructs the graph using a specialized algorithm.
date: 2000.
further information: see <http://acg.media.mit.edu/people/fry/valence/>
further reading: "Organic Information Design" by Ben Fry, unpublished Master's thesis, Media Lab, MIT. <http://acg.media.mit.edu/people/fry/thesis/>

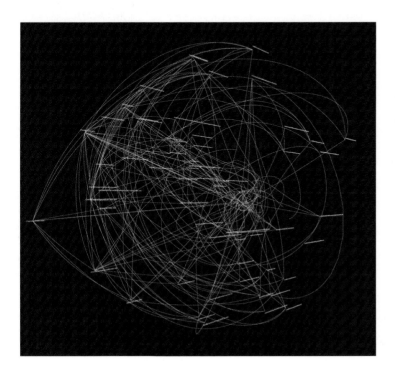

As a continuation of his organic information-design work, Fry has sought to map Web structure in relation to Web usage across a site. His 3-D evolving graphs of site structure and links grow as users traverse a site. As such, as a user explores a new part of the site hierarchy, these are added to the spatialization. The white links are the directory structure and the orange links represent the link structure. The tips of the branches represent individual websites. When a website is visited, the tip thickens to become more visible. This process can be watched, for example viewing the spread of users visiting from an external link as they move through the site. Branches that are not used frequently atrophy and slowly "die", with areas that are not visited eventually removed from the graph. A "movement rule" within the algorithm ensures that a set distance is maintained between individual nodes and their parents. A second rule maintains a distance between nodes and their neighbors to try to minimize overlap. Areas of the structure are labeled and the user can click on nodes to find out which website they relate to.

The images displayed here show the structure and usage of the Aesthetics & Computation Group on the MIT Media Lab's website as it develops over time.

3.20: Anemone

chief cartographer: Ben Fry (Aesthetics & Computation Group, MIT Media Lab).
aim: to produce 3-D dynamic representations of how a site is structured and used.
form: 3-D complex graphs.
technique: monitors site usage from the log and constructs the graph using a specialized algorithm.
date: 2000.
further information: see <http://acg.media.mit.edu/people/fry/anemone/>
further reading: "Organic Information Design" by Ben Fry, unpublished Master's thesis, Media Lab, MIT. <http://acg.media.mit.edu/people/fry/thesis/>

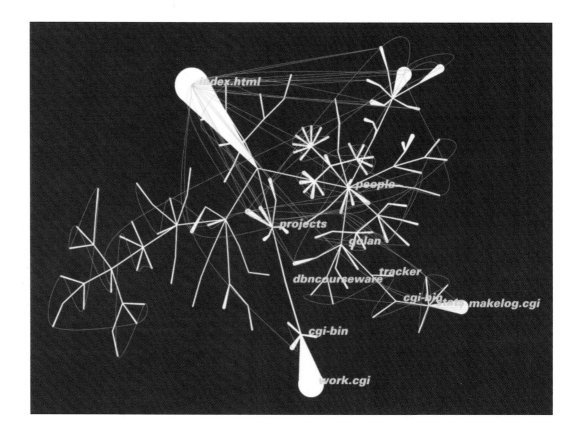

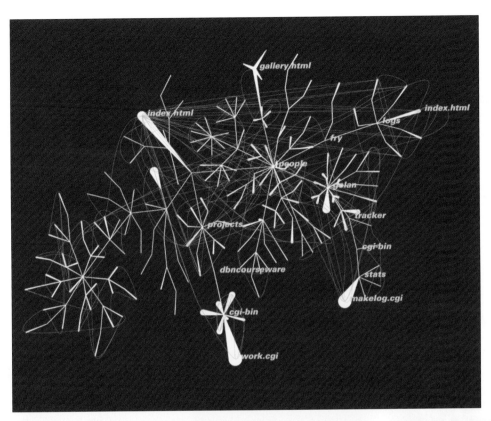

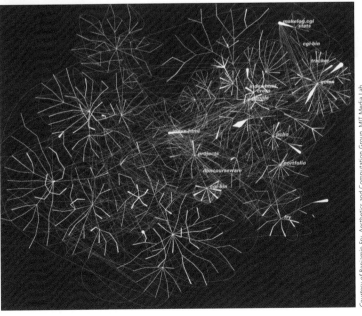

Mapping the web

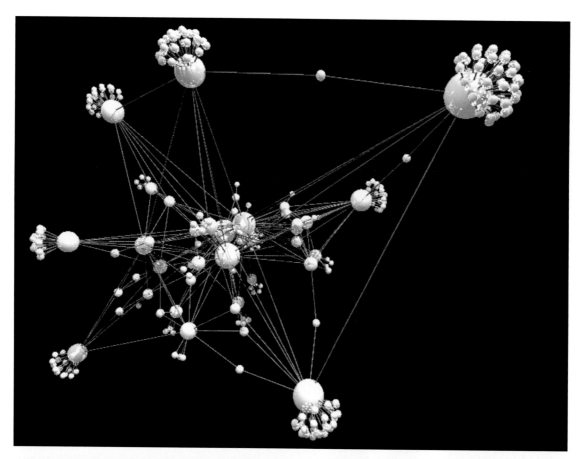

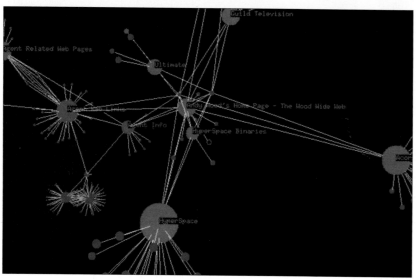

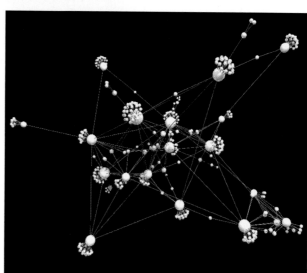

HyperSpace was an early experimental system developed in the mid-1990s to provide a graphical history of browsing trails through the Web. It used 3-D, molecular-style models to display hyperlink-topology structures. Rather than being developed for site managers, HyperSpace was developed in the belief that users could navigate the Web more effectively if they understood how their current location related to the local structure of pages and links. The 3-D models were constructed inside a virtual-reality environment working alongside the user's Web browser, and they were manipulable so that users could inspect the models from any angle. As users moved through the Web and loaded new pages, they were automatically added to the 3-D history graph.

HyperSpace used solid spheres to represent individual Web pages, and arcs for the hyperlinks between them. The size of the sphere was scaled to the number of hyperlinks from the page. The layout of the model in three-dimensional space is achieved using a self-organizing algorithm based on attraction/repulsion behavior of individual lines and spheres – the Web page spheres repulse each other and this is counteracted by the hyperlinks, which attract each other. Starting from a random placement of a clump of interconnected Web pages in the 3-D space, through a series of iterative steps, the spheres and hyperlinks push and pull each other into a stable and coherent spatial arrangement. The result is self-organized equilibrium that provides a distinctive structure of the browsing trail. Only pages that have been explored by the user's browsing are displayed in HyperSpace. The pages at the edge of the explored space are just single nodes, with a single arc back to the parent. This gives the edge spheres a pincushion appearance.

The example images here show various 3-D history graphs formed by a browsing session (the green and white images are ray-traced and not actually generated in real time by the HyperSpace system). One can see how pages that have not yet been explored tend to lie on the edges of the 3-D structure. Unfortunately, HyperSpace did not develop beyond an interesting research prototype.

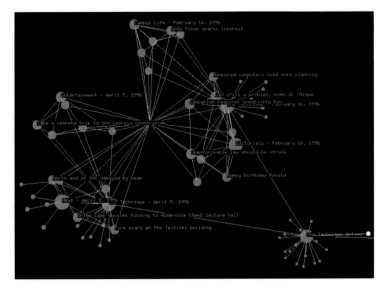

3.21: HyperSpace visualizer

chief cartographers: Andrew Wood, Nick Drew, Russell Beale and Bob Hendley (School of Computer Science, University of Birmingham, UK).
aim: to provide users with a graphic history of their Web surfing.
form: 3-D graph of sphere and lines to represent Web pages and their hyperlink connections. The graph grows as you surf and add new pages.
technique: tracks movement through a site, generating the graphical structure through a self-organizing algorithm.
date: 1995.
further reading: "HyperSpace: Web Browsing with Visualization" by Andrew Wood et al., Third International World-Wide Web Conference Poster Proceedings, 10–14 April 1995, Darmstadt, Germany, pp. 21–25.
<http://www.igd.fhg.de/archive/1995_www95/proceedings/posters/35/index.html>

Like HyperSpace (previous page), WebPath is an example of a "surf map" providing a 3-D graphical history of trails taken by a user through the Web. The angular graph of cubes and pipes representing the browsing structures hangs weightless in a stylized purple cyberworld. It works as a real-time visualization within a virtual-reality environment, working alongside a conventional browser. Individual Web pages are represented by cubes. Cubes were used rather than spheres as their flat surfaces are easier to read from a distance. The page represented by the cubes is indicated by labeling (with the title or URL (universal resource locator) and pasting an image on the faces of the cubes. This image can be the background image of the page, or an image on the page, or background color of the page, depending on the user's choice. Clicking on a cube of interest will load that Web page into the browser.

The positioning of the cubes in the space is used to encode various attributes about the Web page, including when it was accessed. The user can tailor which attributes are active. The three orthogonal dimensions of the space allow one to display three distinct attributes to be used. The vertical axis is used exclusively for the time at which the Web page was accessed so that the cubes at the top of the spatialization are always the most recently visited. The x and y horizontal axes can be used

to encode a variety of attributes, such as loading time of the page, page size or number of hyperlinks, which can be selected by the user. The user can change the meaning of the x and y dimensions at any time, and the cube positions will then be automatically recalculated. WebPath can also position the cube according to approximate real-world geography rather than using an abstract coordinate space. A base map is provided on the "floor" of the information space, and the cubes are positioned in the appropriate country based on the domain name of the website.

The links between Web page cubes show the paths the user has taken via hyperlinks. When a user visits a new Web page, a new cube is created, and a tube connects this back to the previous cube. The color of the tube is used to indicate whether or not the pages are from the same site. Repeat visits to the same website at different times are indicated by multiple cubes that are vertically separated but are connected by solid yellow columns. For a user's most popular Web pages, the column turns from yellow to red to indicate repeated accesses. Distinct browsing sessions are also separated visually using semi-transparent horizontal planes. This divides the space into separate layers.

3.22: WebPath browsing history

chief cartographers: Emmanuel Frécon (Swedish Institute of Computer Science) and Gareth Smith (Lancaster University, UK).
aim: to provide an interactive and tailorable graphic history of Web surfing trails.
form: 3-D cubes and pipes represent pages and the routes traversed by the user. Images on the cube faces are taken from the Web page.
technique: tracks movement through a site, generating the graphical structure through a self-organizing algorithm.
date: 1998.
further information: see <http://www.comp.lancs.ac.uk/computing/users/gbs/webpath/>
further reading: "WebPath – A Three-dimensional Web History" by Emmanuel Frécon and Gareth Smith, IEEE Symposium on Information Visualization (InfoVis '98), 19–20 October 1998, Chapel Hill, N.C, USA.
<http://tina.lancs.ac.uk/computing/users/gbs/ webpath/webpath.html>

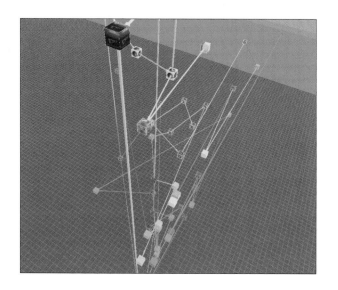

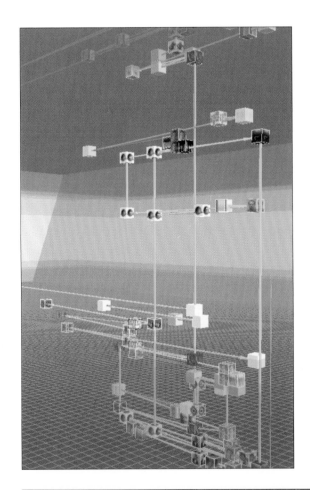

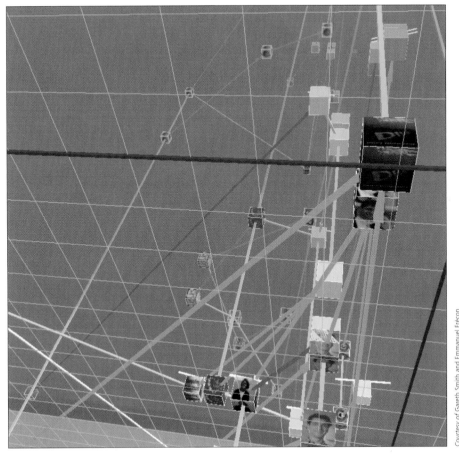

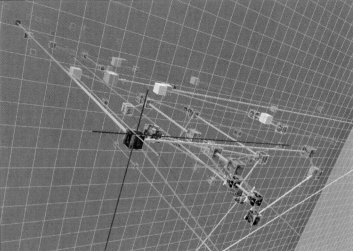

Mapping the web   **113**

"The view from above": 2-D visualization and navigation of the Web

So far in this chapter we have considered attempts to map relatively small sections of the Web. But how do you map the wider Web? An obvious approach is to spatialize the hyperlink structures as some kind of graph, either in two or three dimensions. This is quite easy for individual sites, as we have seen, but to map the many millions of Web pages and their interweaving hyperlinks is an altogether more challenging task. One could argue that it is also a foolish task, given that the Web is in fact its own map. This idea draws on Jorge Luis Borges' famous Cartography fable, recognizing the Web as being both territory and map: by re-mapping it, one simply ends up creating it anew.

While the Web as map and territory is conceptually true, there is clearly a need for new mappings across a range of scales. The Web is now an enormous enterprise consisting of billions of pages. Navigating and searching through this vast information store using simple hypertext "maps" can be difficult, confusing and time-consuming. Site maps and Web management tools seek to address navigation issues at a micro scale. These have been complemented by spatializations that seek to provide a much broader overview detailing the relations across and between large "chunks" of the Web. In general, these types of spatialization can be divided into two broad categories: 2-D and 3-D. In this section, we consider the first set – 2-D spatializations – and examine 3-D spatializations in the following section.

Two-dimensional spatializations adopt a metaphor of the "view from above", of gaining a bird's-eye view. Here it is argued that browsing for a particular piece of information on the Web can often feel like being stuck in an unfamiliar part of town, walking around at street level looking for a particular store. You know the store is around there somewhere, but your viewpoint at ground level is constrained. What you really need is to get above the streets, hovering half a mile or so up in the air, to see the whole neighborhood. In this way you gain important contextual information that allows you to reorientate yourself. This kind of bird's-eye view has been memorably described by David D. Clark, Senior Research Scientist at MIT's Laboratory for Computer Science and the Chairman of the Invisible Worlds Protocol Advisory Board, as the missing "Up button" on the browser. In this section, we examine a number of examples and prototypes for Clark's view of information space.

Yahoo! is the undisputed king of Web directories, providing one of the key information-navigation tools on the Internet. Over several years, it has maintained its popularity as the most-visited website because it does such a good job of shifting, cataloging and organizing the Web. Even so, Yahoo! is a very large site to navigate. ET-Map sought to provide an interactive map of Yahoo! in order to examine whether a map interface, rather than the conventional listing of sites, would be a more useful and effective means of navigating the site. Developed by Hsinchun Chen and the research team in the University of Arizona's Artificial Intelligence (AI) Lab, ET-Map is a hierarchical set of "category maps", which are essentially visual directories.

ET-Map charts a large chunk of Yahoo! from the entertainment section, representing some 110,000 different Web links. The map is a two-dimensional, multilayered category map. Its aim is to provide an intuitive and visual information-browsing tool and to provide the browser with a sense of the lie of the information landscape – what is where, the location of clusters and hotspots, and what is related to what. Ideally, this big-picture all-in-one visual summary needs to fit on a single standard computer screen. ET-Map can be browsed interactively, explored and queried, using the familiar point-and-click navigation style of the Web to find information of interest.

The example shown on page 117 reveals how the spatializations work to reveal information on jazz music. Each "category map" displays groupings of associated Web pages as regularly shaped, homogeneous "subject regions", which can be thought of as virtual "fields" that all contain the same type of information "crop". The spatial extent of the subject regions is directly related to the number of Web pages in that category; for example, the "MUSIC" subject area contains over 11,000 pages and so has a much larger area than the neighboring area of "LIVE", which only has 4,300 or so pages. If a region has more than 200 pages, then a sub-map of greater resolution is created, with a finer degree of categorization. Clicking on a subject region with less than 200 pages takes one to a conventional text listing of the page titles. In addition, a concept of neighborhood proximity is applied, so that subject regions that are closely related in content are plotted close to each other. For example, "FILM" and "YEAR'S OSCARS", at the bottom left of the top layer, are neighbors. The hierarchical nature of the maps, with the ability to "drill down" to different levels, is illustrated over the page.

ET-Map was created using a sophisticated AI technique called Kohonen Self-organizing Map (SOM), which is a neural-network approach that has been used for automatic (i.e., no human supervision) analysis and classification of semantic content of text documents such as Web pages. Chen and his colleagues believe "… that [the] Kohonen SOM-based technique … can be used effectively and scaleably to browse a large information space such as the Internet". However, it is also a challenge to automatically classify pages from a very heterogeneous collection of Web pages, and it is not clear whether the SOM categories will necessarily match the conceptions of a typical user. From the limited usability studies made on category maps, it appears they are good for conducting unstructured, "window shopping" browsing, but are less useful for undertaking more directed searching.

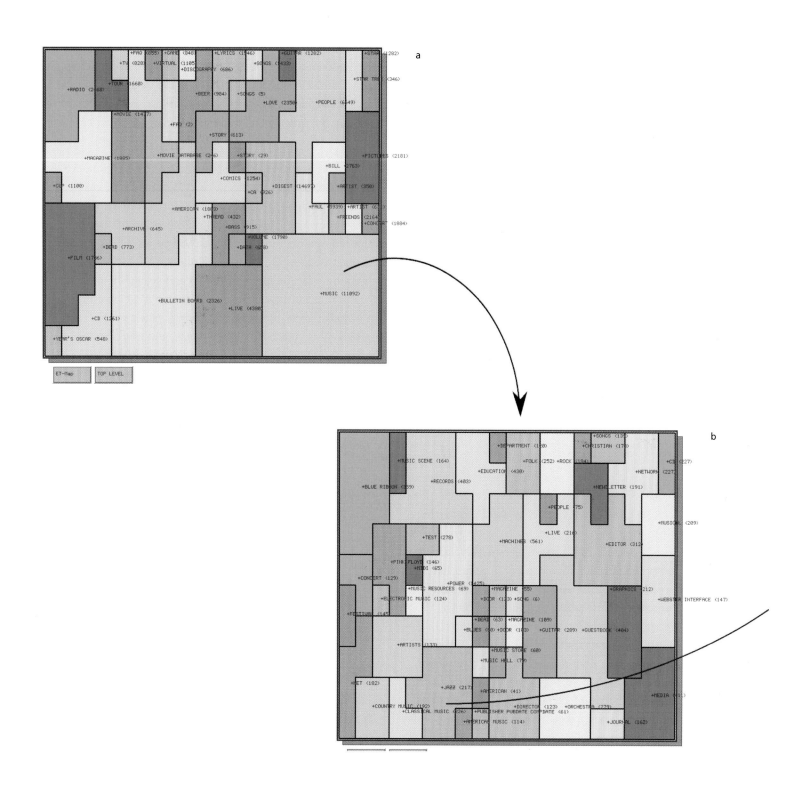

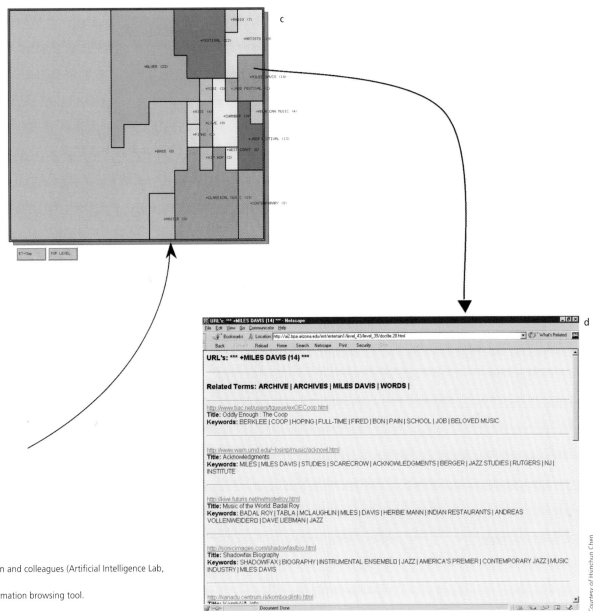

c

d

Courtesy of Hsinchun Chen

3.23: ET-Map

chief cartographer: Hsinchun Chen and colleagues (Artificial Intelligence Lab, University of Arizona, USA).

aim: to provide an intuitive visual information browsing tool.

form: 2-D interactive land-use map.

technique: uses a neural-network approach called Kohonen self-organizing map (SOM) to automatically classify information.

date: 1996–1998.

further information: see <http://ai.bpa.arizona.edu>

further reading: "Internet Categorization and Search: A Self-Organizing Approach" by Hsinchun Chen, Chris Schuffels, and Rich Orwig, 1996, *Journal of Visual Communication and Image Representation*, 7(1) special issue, pp. 88–102.
<http://ai.bpa.arizona.edu/papers/som95/som95.html>

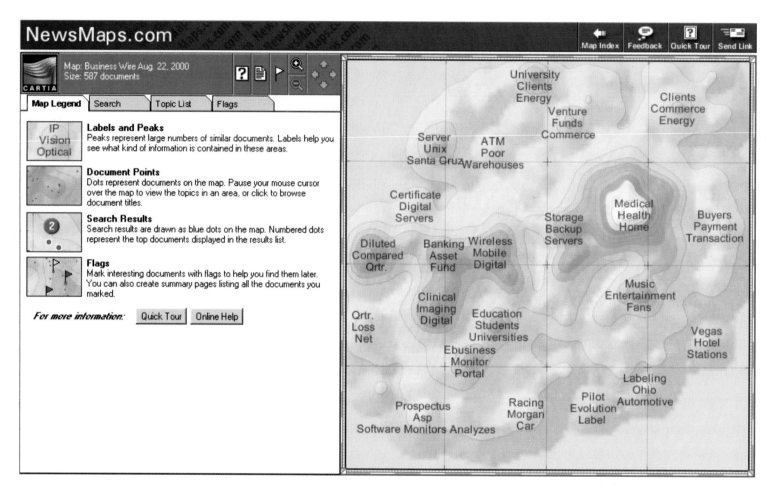

NewsMaps.com

Map Index Feedback Quick Tour Send Link

CARTIA

Map: Business Wire Aug. 22, 2000
Size: 587 documents

Map Legend | Search | Topic List | Flags

Labels and Peaks
Peaks represent large numbers of similar documents. Labels help you see what kind of information is contained in these areas.

Document Points
Dots represent documents on the map. Pause your mouse cursor over the map to view the topics in an area, or click to browse document titles.

Search Results
Search results are drawn as blue dots on the map. Numbered dots represent the top documents displayed in the results list.

Flags
Mark interesting documents with flags to help you find them later. You can also create summary pages listing all the documents you marked.

For more information: Quick Tour   Online Help

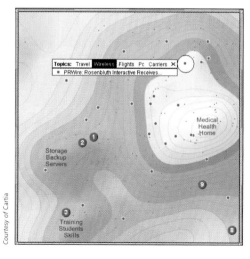

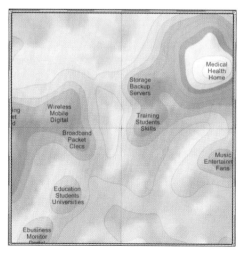

NewsMaps.com has produced one of the best examples of information mapping available on the Web today. It was initially developed as a high-profile "show-and-tell" website to demonstrate the power of ThemeScape information analysis and mapping technology, developed by Cartia Inc. (Based in Bellevue, Washington State, Cartia is a high-tech spin-off formed by information-visualization researchers at Pacific Northwest National Laboratory (PNNL) in 1996.)

NewsMaps' products are among the most map-like of information maps, borrowing literally and liberally from the cartographer's toolbox. The attractive and interactive NewsMaps maps provide a big-picture summary of large volumes of textual information represented as hills, valleys and snow-capped mountain peaks – a cartographic form common on topographical maps of the real world. Here, though, the hills and valleys are used metaphorically to represent the volume of textual information, with peaks representing a large number of news stories discussing the same topic (labeled with keywords). The valleys are the natural transitions between one topic and another. The spatial concept of "neighborhood" is also used, so that the closer together two hills are on the map, the more similar their information content. The actual location of the news articles used to construct the spatialization is indicated by

small black dots. The maps are created using proprietary algorithms and techniques that intelligently summarize the key topics and the relations between them (as with ET-Map – see last example).

A range of NewsMaps is compiled daily to cover international news, US news and technology news. They are delivered over the Web in a fully interactive viewer that allows users to explore each map on their desktop. By passing the mouse cursor over an area of interest, the top five topics within a small radius are displayed in a pop-up window. Clicking once on the terrain will cause a pop-up list of available articles in the area to be displayed. Clicking on an article title of interest allows the full article to be opened in a new browser window. Users can zoom in on a region of the map to see greater local detail, and also to perform a keyword search for articles of interest or select articles from a topic list. The results of such searches or selections are shown prominently by large blue dots on the spatialization, numbered according to their relevance ranking. It is also possible to stick small red marker flags into the terrain to identify documents of interest for future reference and to zoom in and pan around the spatialization to reveal more detail.

3.24: NewsMaps – revealing the topography of Web content

chief cartographer: David Lantrip and colleagues (Cartia, USA).
aim: to show the content of hundreds of online news articles in a single overview map, so as to give users a sense of key stories and themes.
form: two-dimensional landscape of mountains, hills and valleys using shading and contours to give an impression of a terrain. Height of the terrain represents volume of articles.
technique: custom software, known as ThemeScape, which uses proprietary algorithms and techniques that intelligently summarize the key topics and the relations between them.
date: initial research in 1994–96. ThemeScape and NewsMaps were launched in 1998. Screenshots taken in August 2000.
further information: see <http://www.newsmaps.com>; <http://www.cartia.com>
further reading: Original research paper entitled "Visualizing the Non-Visual: Spatial Analysis and Interaction with Information from Text Documents" by J. A. Wise, J. J. Thomas, K. Pennock, D. Lantrip, M. Pottier, A. Schur and V. Crow. Reprinted in *Readings in Information Visualization: Using Vision to Think*, edited by Stuart K. Card, Jock D. Mackinlay and Ben Shneiderman (Morgan Kaufmann Publishers, 1999), pp. 442–450.

Another excellent example of an interactive information map is called "Map of the Market", which aims to display a big-picture summary of the dynamic state of the US stock-market in a single visual snapshot. It does this by mapping the changing stock price and market capitalization of more than 500 major publicly-traded corporations. On a single map, one can quickly gain a sense of the overall market conditions, yet still see many hundreds of individual data elements. This overall picture is very difficult to comprehend from more conventional listings of stock prices. The spatialization was developed by Martin Wattenberg, director of research and development at SmartMoney.com, where he specializes in designing novel ways to visualize financial information.

Map of the Market uses a visualization technique called treemaps to generate the spatializations. This technique was originally developed in the early 1990s by researchers Ben Shneiderman and Brian Johnson at the Human–Computer Interaction Lab at the University of Maryland. Treemaps are a compact way of representing hierarchical data using a space-filling technique of nested, regular tiles. In Map of the Market, the stock market is visualized by a regular quilt of tiles where each tile represents the performance of one corporation (see top-right map). Three key visual components in the map – tile size, color and position – are used to encode attributes of the corporation. First, the tile size is scaled to the market capitalization of the company, so that the bigger the tile, the more valuable the company in stock-market valuation. The color, of the tile represents the percentage movement in a company's stock price. The color scheme used runs from green to red and is familiar in financial markets. The brighter the color the bigger the change (either positive or negative) in stock price. The tile is shaded black if there is no change in price. One can interactively choose the time period over which the stock-price change is calculated (from the last market close, to six months or a year).

Company tiles are grouped into familiar hierarchies based on classifications of industries (e.g., software, networking, semiconductors) which in turn form broader sectors (e.g., technology, energy, financial). Within these groups the spatial arrangement of individual tiles is organized using a neighborhood technique that places similar companies near each other. In this information map, similarity is a metric based on historically similar stock-price performance.

Map of the Market also provides many useful interactive features. Information relating to individual companies can be viewed by brushing the mouse cursor over the tiles (see bottom-left). This causes a small pop-up box to appear containing summary data. Clicking on this box provides access to detailed background news and information from other pages at the SmartMoney website. It is also possible to zoom into the map to focus on a particular sector and industry of interest. For example, in the screenshots opposite, we have clicked on the large green square tile representing the networking company Intel (bottom-left). Then we zoomed into the sector map for technology, containing seven different industries (bottom-middle). Finally, we zoomed into the software industry map which is dominated by the large tiles of Microsoft and Oracle (bottom-right). Other interactive features include the ability to change the color-coding, to turn on and off the labeling of news headlines associated with particular companies, and to highlight the top five winners and losers.

SmartMoney is also developing further implementations of its information mapping technology to different financial information. For example its Mutual Fund map, shown top-right, visualizes the performance of 1,000 mutual funds in the United States. The example screenshot was taken in August 2000 and uses the alternative blue–yellow color scheme. It is also possible to map a personalized stock portfolio once registered as a SmartMoney user.

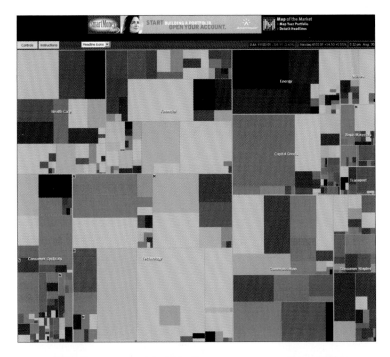

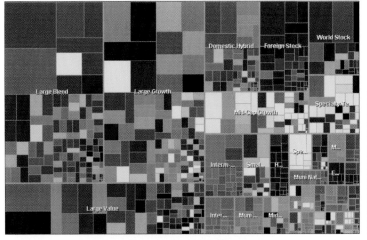

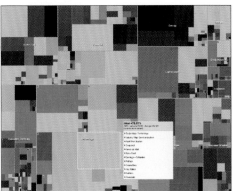

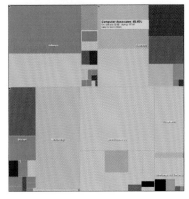

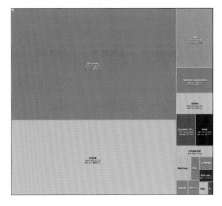

3.25: Mapping the money in cyberspace – "Map of the Market"

chief cartographer: Martin Wattenberg (SmartMoney.com, New York).

aim: to show the performance of a stock-market in a single visual snapshot by mapping the capitalization and changing stock price of more than 500 companies.

form: an information quilt using tiles to represent companies, with size, color and position encoding key attributes.

technique: interactive Java applet. Metaphor is based on a technique called treemaps.

date: Launched at the end of 1998. Screenshots taken August 2000.

further information: see <http://www.smartmoney.com>

further reading: "Treemaps for space-constrained visualization of hierarchies", by Ben Shneiderman for Human–Computer Interaction Lab, 8 November 2000.
<http://www.cs.umd.edu/hcil/treemaps/>.

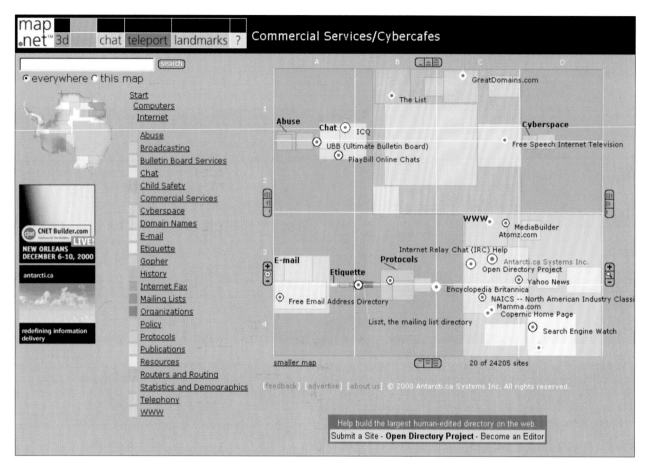

search
● everywhere ○ this map

Start
Computers
Internet

Abuse
Broadcasting
Bulletin Board Services
Chat
Child Safety
Commercial Services
Cyberspace
Domain Names
E-mail
Etiquette
Gopher
History
Internet Fax
Mailing Lists
Organizations
Policy
Protocols
Publications
Resources
Routers and Routing
Statistics and Demographics
Telephony
WWW

CNET Builder.com
NEW ORLEANS
DECEMBER 6–10, 2000
LIVE!
antarcti.ca
redefining information delivery

GreatDomains.com
The List
Abuse
Chat ⊙ ICQ
UBB (Ultimate Bulletin Board)
PlayBill Online Chats
Cyberspace
Free Speech Internet Television

WWW ⊙ MediaBuilder
Atomz.com

Internet Relay Chat (IRC) Help
E-mail **Protocols** Antarcti.ca Systems Inc.
Open Directory Project
Etiquette Yahoo News
Encyclopedia Britannica
Free Email Address Directory NAICS -- North American Industry Classi
Mamma.com
Copernic Home Page
Liszt, the mailing list directory Search Engine Watch

smaller map 20 of 24205 sites

[feedback] [advertise] [about us] © 2000 Antarcti.ca Systems Inc. All rights reserved.

Help build the largest human-edited directory on the web.
Submit a Site - **Open Directory Project** - Become an Editor

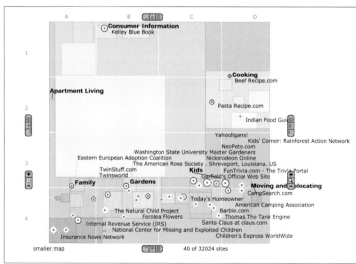

A B C D
Consumer Information
Kelley Blue Book

⊙**Cooking**
Beef Recipe.com

Apartment Living
⊙ Pasta Recipe.com

Indian Food Gui

Yahooligans!
Kids' Corner: Rainforest Action Network
NeoPets.com
Washington State University Master Gardeners
Eastern European Adoption Coalition Nickelodeon Online
The American Rose Society . Shreveport, Louisiana. US
TwinStuff.com FunTrivia.com - The Trivia Portal
Twinsworld **Kids** Garfield's Official Web Site
⊙**Family** **Gardens** **Moving and Relocating**
CampSearch.com
Today's Homeowner
American Camping Association
The Natural Child Project Barbie.com
Fernlea Flowers Thomas The Tank Engine
Internal Revenue Service (IRS) Santa Claus at claus.com
National Center for Missing and Exploited Children
Insurance News Network Children's Express WorldWide
smaller map 40 of 32024 sites

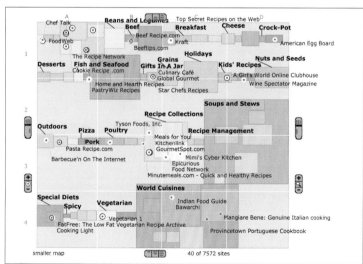

A B C D
Chef Talk **Beans and Legumes** Top Secret Recipes on the Web
Beef **Breakfast** **Cheese** **Crock-Pot**
⊙ FoodWeb Beef Recipe.com
Kraft American Egg Board
Beeftips.com
The Recipe Network **Holidays**
Desserts **Fish and Seafood** **Grains** **Nuts and Seeds**
Cookie Recipe .com **Gifts In A Jar** **Kids' Recipes**
Culinary Café A Girl's World Online Clubhouse
Global Gourmet Wine Spectator Magazine
Home and Hearth Recipes
PastryWiz Recipes Star Chefs Recipes

Soups and Stews
Recipe Collections
Tyson Foods, Inc.
Outdoors **Pizza** **Poultry** **Recipe Management**
Meals for You!
Pork Kitchenlink
Pasta Recipe.com GourmetSpot.com Mimi's Cyber Kitchen
Barbecue'n On The Internet Epicurious
Food Network
Minutemeals.com - Quick and Healthy Recipes
World Cuisines
Special Diets Indian Food Guide
Spicy **Vegetarian** Bawarchi
Vegetarian 1 Mangiare Bene: Genuine Italian cooking
FatFree: The Low Fat Vegetarian Recipe Archive
Cooking Light Provincetown Portuguese Cookbook
smaller map 40 of 7572 sites

Courtesy of Map.net

122 Atlas of cyberspace

Map.net is a novel Web-search directory that employs perhaps the most comprehensive system of information spatialization currently available on the Web. The directory uses both 2-D multilevel information maps and 3-D fly-through cityscapes to spatialize over two million websites from the Open Directory (a well-known information directory available on the Internet). Map.net was launched in November 2000 and its underlying spatialization technology is called Visual Net. It is the brainchild of Tim Bray, who is not only an expert in hypertext and information navigation but also the co-creator of the XML standard. His company Antarcti.ca's self-stated aim is to "transform networks into places" and it believes information spatialization is the best route to achieve this.

Map.net spatializes the two million or so websites in existence onto the land surface of the continent of Antarctica. They are grouped in a hierarchy of categories that are represented visually on the map display as nested rectangular tiles. The classification of websites into categories is achieved by the 30,000 or so volunteer editors at the Open Directory. The rectangular category tiles are color-coded and their size is in proportion to the number of websites they contain. For example, in the screenshot shown top-opposite we see the tiles for the Internet-related categories.

The legend to the left of the map lists the categories in order with their color coding. The small map of Antarctica in the top left-hand corner of the screen shows the location of this information map in relation to the rest of the Web mapped onto the land surface. Map.net's 2-D information maps have many common features with ET-Map (see page 116–17) in terms of information representation, except that the categorization is derived by humans rather than through any automatic classification of websites. Also, the tile arrangement is not based on similarity of content but is simply laid out in alphabetical order from top-left to bottom-right in each image. The maps are fully interactive, allowing the user to move up and down the hierarchy of maps (effectively an informational zoom). Using the small control buttons around the edge of the map allows the user to pan around to see neighboring regions at the same scale. Clicking on a tile of interest will usually provide a more detailed map of just that category (see below).

Significantly, the actual position and characteristics of individual websites are shown on the maps as well. These are represented by the distinctive circular targets, and clicking on one will open the site in a new browser window. By default, the top-20 most visible websites in a particular map region are shown, but this parameter can be altered by the user. Visibility is an important metric computed by Map.net and is represented by an overall score to identify the "best" websites in the vast milieu of the Web. The visibility score is made up of four components and these are encoded graphically in the target symbols. The thickness of the outer black ring indicates how many outgoing hyperlinks the site has to other pages on the Web, while the thickness of the inner white ring is a measure of the site's popularity based on the number of incoming hyperlinks from

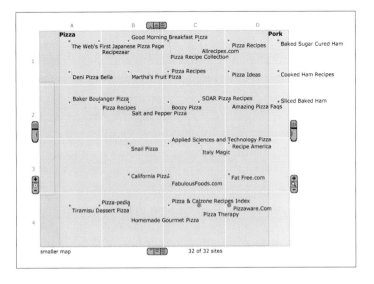

3.26: Map.net 2-D multilevel information maps
chief cartographer: Tim Bray (Antarcti.ca).
aim: to provide a visual Internet search directory.
form: nested, rectangular, color-coded tiles and websites shown by red and white targets.
technique: Visual Net technology spatializes the Open Directory.
date: launched November 2000. Screenshots taken late November 2000.
further information: see <http://map.net>

the rest of the Web. The red circle at the center of the target symbol shows the relative size of the site in terms of the number of pages it has. Lastly, the small yellow arrows added to the outside of the symbol indicate that it has a "cool" rating from one of the editors at Open Directory.

This interactive, multilevel information map enables several million websites to be spatialized and browsed. As an example, the spatializations display a typical browsing sequence through three layers. As we navigate deeper into the information hierarchy, we see more detailed categories and the maps show much finer granularity of information. To begin, we have a general map of "Home" (bottom-left), one of the 16 top-level categories. The "Home" category is represented by the large aquamarine rectangle that fills the whole map window. It contains 17 subcategories, which are shown by the variously sized and colored rectangles. These are arranged in alphabetical order from the tiny "Apartment Living", followed by the huge "Consumer Information", right down to "Urban Living". The most visible 40 websites, from a total of 32,024, are shown by the target symbols, with the majority of these coming from the "Kids" category.

Say we wanted to know more about cooking. At this scale, only three individual websites are shown, but clicking on the "Cooking" tile zooms in to reveal a more detailed map of this category (bottom-right). This is a more complex-looking map, with greater information granularity. Many more category tiles are shown, summarizing over 7,500 individual websites. If we are particularly interested in pizza, say, we can drill down further by clicking on the "Pizza" tile. This takes us to the most detailed available map, at the bottom of the hierarchy of categories (see map on p. 123). On this map all 32 pizza-related websites are shown. Most are just represented by a red dot with very narrow black or white rings, revealing that they are not well-linked or visible. None of them has been rated "cool". Clicking on the background on this map, because there are no further levels, will load the 3-D fly-through cityscape view of this category.

Real-world geography and conventional cartography can be used to provide a spatialization of websites to aid navigation. These types of map are commonly known as "sensitive maps" or "clickable maps", and one of the best examples is the UK Academic Map produced and maintained by Peter Burden of the University of Wolverhampton. The UK Academic Map from summer 2000 is shown on page 127. The original went online in July 1994 – ancient history for the Web! – and this is version 5.

In the early days of the Web in 1994 and 1995, sensitive maps like this one were a quite popular way of cataloging websites in a region or country (where another good example was the Virtual Tourist site). However, this approach did not scale well to keep pace with the tremendous growth in the number and type of websites. The UK Academic Map is one of the few examples that is still maintained and useful. Arguably, this is because it has specialized in cataloging a very limited segment of websites – those of UK academia – which is relatively static in size and Web address. The current version maps 204 different institutions (and other maps are provided showing colleges and research centers).

The map interface comprises a basic coastline of the UK onto which the university websites are plotted and labeled. The position of the symbols is as close as possible to the actual geographic location of the university. Differently colored and shaped symbols are used to denote different types of institution. Using the map as an information navigation tool is very easy: one simply locates the university of interest and clicks on its marker on the map to open that institution's homepage. One can choose a particular access point into the university's Web page, such as the prospectus or alumni page. For example, if one wanted to access the website of the University of Plymouth one would simply click on the red star symbol located on the city and its homepage would open. Or if one was interested in the students' union at the University of Kent one would just select that topic from the list and then click on the star located at Canterbury and again the appropriate page would open. In this fashion, the map is said to be "clickable" or "sensitive" in that it can sense the position of mouse clicks and respond. As such, even though this geographic directory of websites is inherently static, in that the map has to be manually updated, it offers limited interaction to the end user.

Clearly, there are some limitations with this kind of geographic map-based Web-directory approach, both in terms of graphical form and also at the conceptual level of information navigation. In graphical terms, data can quickly become cluttered as many websites tend to cluster in urban areas (see chapter 2). This is apparent with the high density of universities in Manchester and London making the designer's job difficult in placing all the symbols and, particularly, their labels. In response, the map's creators have been forced to abandon the geographic representation in these circumstances and simply present them as a long list on the right-hand side of the map, floating in the North Sea. For certain other universities that could not be fitted close to their geographic location, the authors have had to use long guiding lines to link the town to the symbol and label drawn on the edge of the map. In many ways these cartographic "fudges" are necessary in order to portray a volume of information that does not easily fit into the constrained bounds of geographic space. One obvious solution would be to break the UK up into smaller geographic units and have separate, small-scale maps for each. This would effectively provide greater expanse of "screen space" to represent the information and may increase legibility, but it would also impair the usability for information navigation. In particular, it would be much harder to browse, and the reader would lose the opportunity to see the "big picture".

Another, more conceptual weakness of this type of navigation is that it is not appropriate for many kinds of resources on the Web. In our example of universities in the United Kingdom, it is generally suitable because these types of institution are strongly associated with a city or town; indeed, many are named after the place in which they are located. However, for many other types of information, a geographic location may not be an appropriate key for an index. Another important issue with using geography as the index is that a lot of people have a weak sense of geography in terms of finding cities or countries on a map. Using geography as the key may therefore make information navigation harder compared with, say, browsing through an alphabetic list. For example, if one wanted to find the University of Plymouth using the geographic directory and one did not know where the city of Plymouth was located in the UK, it would take a while to visually search the whole map.

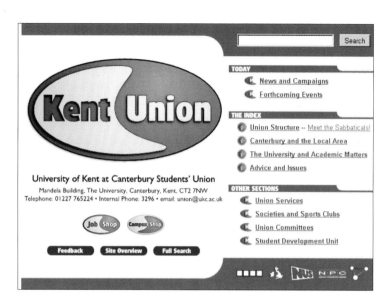

3.27: Geographic map as Web directory – UK Academic Map

chief cartographer: Peter Burden (School of Computing and Information Technology, University of Wolverhampton, UK).

aim: a geographic index of all UK universities on a single map, to provide an easy way to access their websites.

form: a simple outline map of the UK with universities located by graphic markers, along with text labels of each institution's name (see following page).

technique: simple HTML image map.

date: launched in July 1994; currently version 5 as at September 1998.

further information: see <http://www.scit.wlv.ac.uk/ukinfo/uk.map.html>

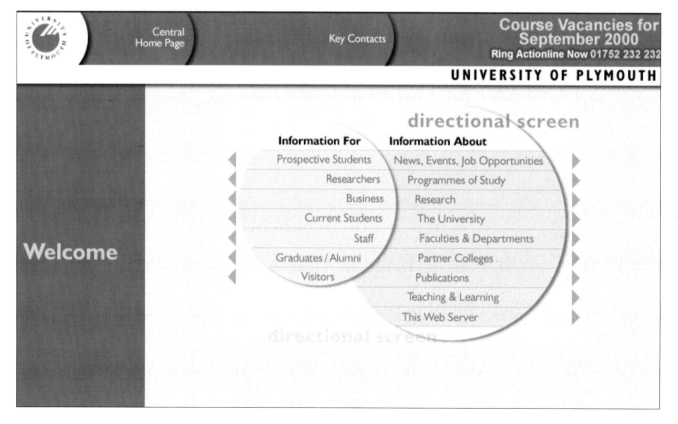

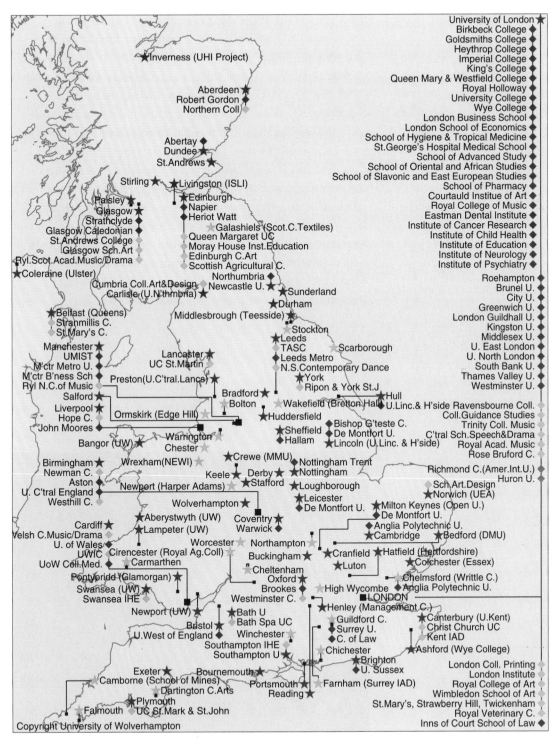

Inverness (UHI Project)

Aberdeen
Robert Gordon
Northern Coll

University of London
Birkbeck College
Goldsmiths College
Heythrop College
Imperial College
King's College
Queen Mary & Westfield College
Royal Holloway
University College
Wye College
London Business School
London School of Economics
School of Hygiene & Tropical Medicine
St.George's Hospital Medical School
School of Advanced Study
School of Oriental and African Studies
School of Slavonic and East European Studies
School of Pharmacy
Courtauld Institue of Art
Royal College of Music
Eastman Dental Institute
Institute of Cancer Research
Institute of Child Health
Institute of Education
Institute of Neurology
Institute of Psychiatry

Abertay
Dundee
St.Andrews
Stirling   Livingston (ISLI)
Edinburgh
Napier
Heriot Watt
Galashiels (Scot.C.Textiles)
Queen Margaret UC
Moray House Inst.Education
Edinburgh C.Art
Scottish Agricultural C.

Paisley
Glasgow
Strathclyde
Glasgow Caledonian
St.Andrews College
Glasgow Sch.Art
Ryl.Scot.Acad.Music/Drama
Coleraine (Ulster)

Roehampton
Brunel U.
City U.
Greenwich U.
London Guildhall U.
Kingston U.
Middlesex U.
U. East London
U. North London
South Bank U.
Thames Valley U.
Westminster U.

Cumbria Coll.Art&Design
Carlisle (U.N'thmbria)
Northumbria
Newcastle U.
Sunderland
Durham
Middlesbrough (Teesside)
Stockton
Leeds
TASC   Scarborough
Leeds Metro
N.S.Contemporary Dance
York
Ripon & York St.J
Hull
U.Linc.& H'side Ravensbourne Coll.
Coll.Guidance Studies
Trinity Coll. Music
C'tral Sch.Speech&Drama
Royal Acad. Music
Rose Bruford C.
Richmond C.(Amer.Int.U.)
Huron U.

Belfast (Queens)
Stranmillis C.
St.Mary's C.
Manchester
UMIST
M'ctr Metro U.
M'ctr B'ness Sch
Ryl N.C.of Music
Salford
Liverpool
Hope C.
John Moores

Lancaster
UC St.Martin

Preston(U.C'tral.Lancs)
Bradford
Bolton   Wakefield (Bretton Hall)
Ormskirk (Edge Hill)
Huddersfield
Warrington
Chester
Sheffield   Bishop G'teste C.
Hallam   De Montfort U.
Lincoln (U.Linc. & H'side)

Bangor (UW)
Crewe (MMU)
Wrexham(NEWI)
Nottingham Trent
Nottingham
Keele   Derby
Stafford
Loughborough
Leicester
De Montfort U.

Birmingham
Newman C.
Aston
U. C'tral England
Westhill C.

Newport (Harper Adams)
Wolverhampton

Sch.Art.Design
Norwich (UEA)
Milton Keynes (Open U.)
De Montfort U.
Anglia Polytechnic U.
Cambridge   Bedford (DMU)

Cardiff
Velsh C.Music/Drama
U. of Wales
UWIC
UoW Coll.Med.
Pontypridd (Glamorgan)
Swansea (UW)
Swansea IHE

Aberystwyth (UW)
Lampeter (UW)
Coventry
Warwick

Worcester   Northampton
Cirencester (Royal Ag.Coll)
Buckingham
Carmarthen

Cheltenham
Oxford
Brookes
Westminster C.

Cranfield   Hatfield (Hertfordshire)
Luton   Colchester (Essex)
Chelmsford (Writtle C.)
High Wycombe   Anglia Polytechnic U.

Newport (UW)
Bath U
Bath Spa UC
Bristol
U.West of England
Winchester
Southampton IHE
Southampton U

LONDON
Henley (Management C.)
Guildford C.   Canterbury (U.Kent)
Surrey U.   Christ Church UC
C. of Law   Kent IAD
Chichester   Ashford (Wye College)
Brighton
U. Sussex

Exeter   Bournemouth
Camborne (School of Mines)
Dartington C.Arts
Portsmouth   Farnham (Surrey IAD)
Reading
Plymouth
Falmouth   UC St.Mark & St.John

London Coll. Printing
London Institute
Royal College of Art
Wimbledon School of Art
St.Mary's, Strawberry Hill, Twickenham
Royal Veterinary C.
Inns of Court School of Law

Copyright University of Wolverhampton

Spiral and STARRYNIGHT are interactive interfaces for browsing a large online archive of articles about net art at Rhizome. Both are clearly inspired by astronomy, employing the metaphor of stars floating in an infinite black void to represent individual articles. Rhizome is a not-for-profit online forum for the presentation and discussion of new media art. Its discussion list is edited and indexed into an archive of several thousand articles from the past several years. Each star in the STARRYNIGHT interface represents a single article in the archive. Each time an article gets selected (and read) the star gets a little bit brighter. Over time, then, the interface encodes the popularity of different articles. The position of the star in the interface is random, and new stars are added on a weekly basis.

Browsing STARRYNIGHT consists of passing the mouse cursor over the stars, which causes article keywords to "pop up". Selecting a keyword of interest leads to the display of a constellation showing other similar stars (articles). Finally, clicking on a star causes the article to be opened in the browser. According to the site, "STARRYNIGHT is both a mirror and a map. On the one hand, it offers a reflection of the Rhizome community's reading habits . . . On the other hand, it acts as a navigational interface by connecting similar stars/texts into constellations regardless of their brightness."

Spiral uses many of the same metaphorical aspects as are employed in STARRYNIGHT except that it highlights the time dimension. Like STARRYNIGHT, Spiral is an interactive interface providing a view from "above" a Spiral galaxy of stars (bottom-left). The position of the stars along the spiral arm is chronological, so that the oldest stars are at the "center" of the galaxy. Using the slider on the right-hand side of the display, a user can smoothly rotate the spiral galaxy, thereby scrolling backward and forward through time. "Brushing" the mouse cursor over the stars reveals the summary details about the article. Clicking on a star of interest loads the full article into a semi-transparent pane on top of the spiral display (as can be seen bottom-right). Within the plane of the spiral, different strands of stars represent different broad categories of articles(such as events, reviews theory). One can access the archive using conventional keyword searching and browsing.

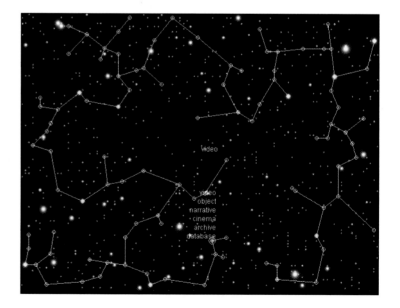

3.28: Artistic searching/searching art – STARRYNIGHT and Spiral

artist: STARRYNIGHT by Alex Galloway and Mark Tribe with Martin Wattenberg;
Spiral by Martin Wattenberg with Steve Cannon (Rhizome).
aim: to provide a visual interface to browse a large archive of articles.
form: astronomic metaphor, with articles represented as stars.
technique: interactive browsing interfaces. Brightness of star icons relates to article popularity. STARRYNIGHT shows virtual constellations of related articles based on indexed keywords; Spiral employs interactive animation to display the stars in chronological order.
date: STARRYNIGHT – 1999; Spiral – 2000. Screenshots taken in September 2000.
further information: see <http://www.rhizome.org>

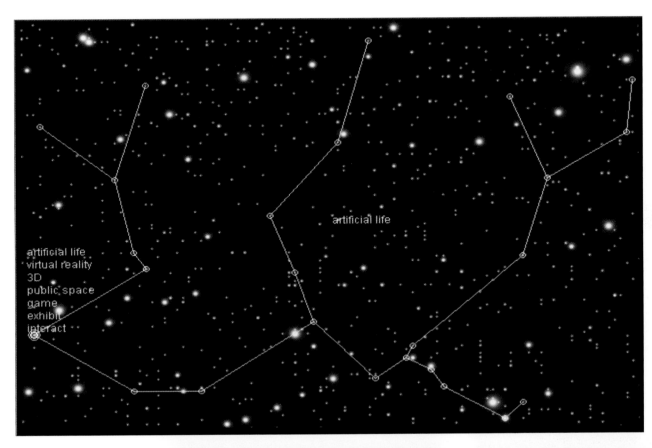

artificial life

artificial life
virtual reality
3D
public space
game
exhibit
interact

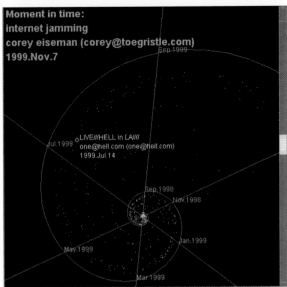

Moment in time:
internet jamming
corey eiseman (corey@toegristle.com)
1999.Nov.7

Sep.1999

LIVE///HELL in LA////
one@hell.com (one@hell.com)
1999.Jul.14

Jul.1999

Sep.1998

Nov.1998

Jan.1999

May.1999

Mar 1999

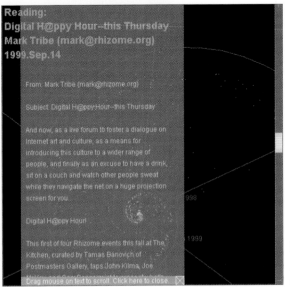

Reading:
Digital H@ppy Hour--this Thursday
Mark Tribe (mark@rhizome.org)
1999.Sep.14

From: Mark Tribe (mark@rhizome.org)

Subject: Digital H@ppy Hour--this Thursday

And now, as a live forum to foster a dialogue on
internet art and culture, as a means for
introducing this culture to a wider range of
people, and finally as an excuse to have a drink,
sit on a couch and watch other people sweat
while they navigate the net on a huge projection
screen for you...

Digital H@ppy Hour!

This first of four Rhizome events this fall at The
Kitchen, curated by Tamas Banovich of
Postmasters Gallery, taps John Klima, Joe

Drag mouse on text to scroll. Click here to close.

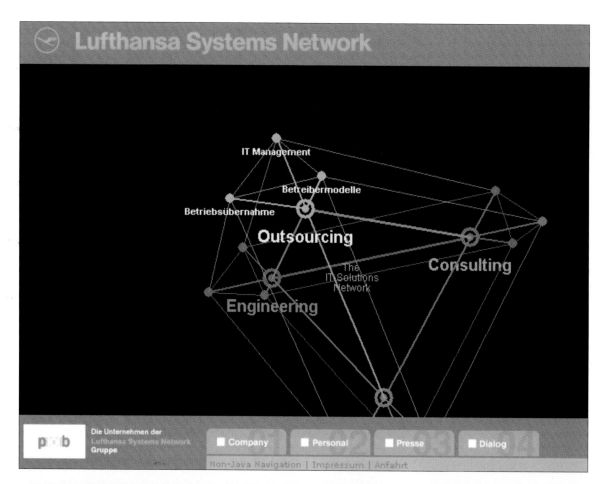

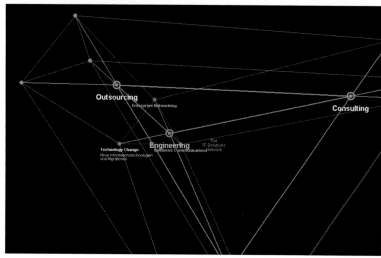

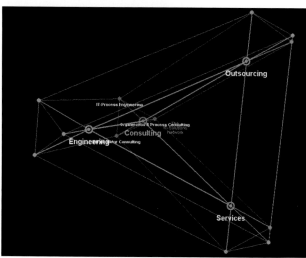

"The view from within": 3-D visualization and navigation of the Web

In the final section of this chapter, we consider mappings that extend methods to visualize and navigate the Web into three dimensions. Here, rather than seeking a "view from above", the spatializations seek to place the user in the heart of the data: "the view from within". In many ways, these spatializations are the closest in nature to those envisaged by William Gibson in his descriptions of cyberspace (see chapter 5).

As with other sections, we order the discussion along a scale continuum from the local to the global, with those concerned with the same scale then ordered from simple to complex.

Fork's intriguing 3-D network is very much at the boundary between a practical navigation tool and a work of art, being created by hand rather than by algorithm. Clearly, this divide is a fuzzy one, as we show in chapter 5. The creators envision Web navigation using a triangular orange graph gently floating and spinning in blank space. The 3-D network metaphor in many ways matches the nature of the organization and website being mapped – it being the various IT-related departments of the German airline Lufthansa, this part of the company being called the "Systems Network". The intersections of the triangles in the graph form nodal points representing different departments. Users can freely interact with the graph, rotating it and zooming into nodes to reveal more details. Clicking on the labels takes one to the appropriate Web page.

3.29: 3-D navigation interface for Lufthansa Systems Network

designer: Fork Unstable Media for Lufthansa Systems Network.
aim: to show the structure of IT departments in the Systems Network group of Lufthansa and enable users to navigate their website.
form: 3-D triangular graph floating in space. The user can rotate the graph and zoom in to see more detail. Clicking on a title opens the relevant Web page.
technique: 3-D interactive graphics using Java.
date: 1999. Screenshots taken in October 2000.
further information: see <http://www.lhsysnet.com> and <http://www.fork.de>
further reading: *Browser 2.0: The Internet Design Project* by Patrick Burgoyne and Liz Faber (Lawrence King Publishing, London, 1999).

A world of words forever swirling and shifting is the vision of cyberspace presented by the Visual Thesaurus. It was created in 1998 by Plumb Design, an innovative New York-based design firm, in order to demonstrate the potential of its Thinkmap visualization technology. The company claims that this product is "an engaging experience in language and interface . . . an artistic exploration that is also a learning tool. Through its dynamic interface, the Visual Thesaurus alters our relationship with language, creating poetry through user action, dynamic typography and design."

Visual Thesaurus provides an artistic interface to surf and explore the WordNet thesaurus of over 150,000 words, phrases and meanings, as well as the multitudes of associations and dependencies. (WordNet itself is a free-of-charge digital thesaurus database developed by the Cognitive Science Laboratory at Princeton University.) When Visual Thesaurus is initiated, the words are presented in a sterile white 3-D space. Clicking on words adds connected words, creating a complex and evolving linguistic mesh. The strength of linguistic relationships between words and phrases is indicated by distance and brightness in the display. Users can navigate through the thesaurus by clicking on words in the 3-D space or by entering them into the search box at the bottom of the display. Users can change how the thesaurus retrieves words, for example focussing on verbs or nouns.

Although interesting, it can be disorientating to use this product and in many ways it is more an experience than a lexical tool.

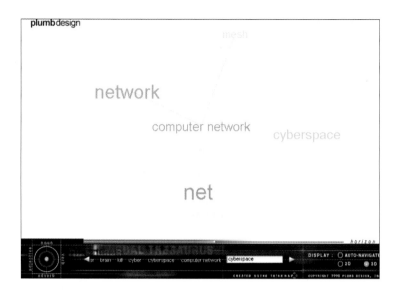

3.30: Visual Thesaurus

designer: Plumb Design, Inc. (New York).
aim: to demonstrate the power of Thinkmap software for providing interactive displays of complex information. This demo visualizes a large thesaurus.
form: a continually swirling 3D graph of words, with lines between them showing linguistic relationships.
technique: 3-D interactive graphics using a custom Java application.
date: 1998. Screenshots taken in October 2000.
further information: see <http://www.plumbdesign.com> and <http://www.thinkmap.com>
further reading: "Choose a Word and Float an Idea", by Matthew Mirapaul, *New York Times*, 12 March 1998. <http://www.nytimes.com/library/tech/98/03/cyber/artsatlarge/12artsatlarge.html>

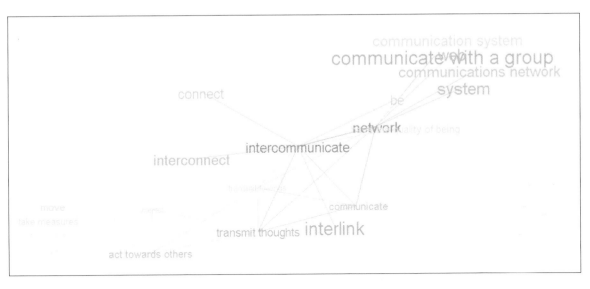

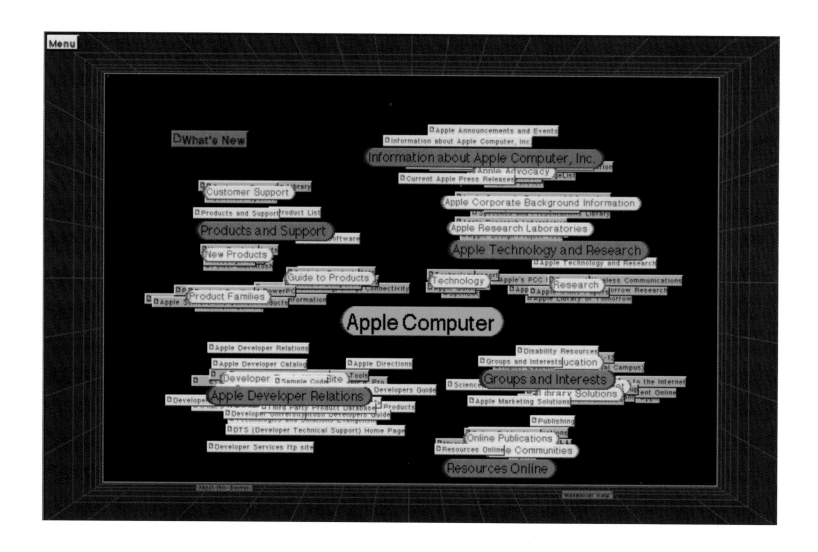

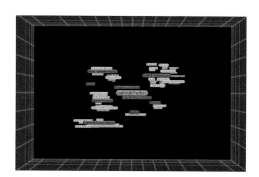  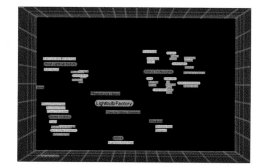

HotSauce was an interesting and largely experimental three-dimensional fly-through interface for navigating information spaces such as websites or the files on a PC disk. It was developed – virtually as a one-man effort – by Ramanathan V. Guha in the mid-1990s while he was at Apple Research. The basic concept behind HotSauce can be expressed by the phrase "Why just browse when you can fly?", the idea being that an immersive 3-D interface to spatialize information would aid navigation. The underlying format to structure data into hierarchies for specific spatialization was called Meta-Content Framework (MCF), also developed by Guha.

HotSauce worked as a browser plug-in, so that when you went to an MCF website you were dropped into a first-person perspective view of the information space, with pages floating as brightly colored asteroid-like blocks in an infinite black space. The view is somewhat like that from a "starship cockpit", and you smoothly fly through the landscape with page blocks becoming larger as they are approached and then disappearing as they are "passed". Web pages are represented by rectangular blocks that are labeled with the page title. Broader "topics" are indicated by round-cornered rectangles, and these provide organizing structure to the information space. Different hierarchical levels of the information space are denoted by different colors of the floating blocks, as well as their spatial depth in the 3-D display. It is easy to fly into and around the space using the mouse to guide the direction of flight and holding down buttons to go forward and backward. A page can be accessed by simply double-clicking on the relevant block.

The image top-left on p. 134 shows the default HotSauce view of the Apple website, circa 1997, with the main green "Apple Computer" topic at the front, which is then followed by major sections of the site represented by the bold red blocks. Further back still are yellow and then purple pages. The remaining screenshots try to capture the essence of the 3-D fly-through effect by showing a sequence for the Lightbulb Factory website.

Although smoothly zooming through HotSauce space is quite fun, it is surprisingly hard to find pages and fly toward them. Once immersed in the space and surrounded by blocks, it is easy to become disorientated. Despite the practical difficulties in actually using HotSauce for information browsing and retrieval, it remains an interesting experiment in mapping information. Unfortunately, Apple ended its development of HotSauce in 1997 and Guha moved on to other projects.

3.31: HotSauce information fly-through

chief cartographer: Ramanathan V. Guha (while working at Apple Research).
aim: to provide a way to browse information spaces with a 3-D interface.
form: fly-through interface, with Web pages represented as brightly colored rectangles floating in an infinite black space.
technique: small plug-in for Web browsers. Built on top of the Meta-Content Framework (MCF) for organizing and describing information spaces.
date: developed between June 1995 and April 1997 and publicly released toward the end of 1996. The screenshots here were taken in August 2000.
further information: R.V. Guha's homepage at <http://www.guha.com/>, while the HotSauce plug-in and documentation are still available from the Lightbulb Factory at <http://www.xspace.net/hotsauce/>
further reading: "Innovators of the Net: Ramanathan V. Guha and RDF" by Marc Andreessen, 8 January 1999.
<http://home.netscape.com/columns/techvision/innovators_rg.html>

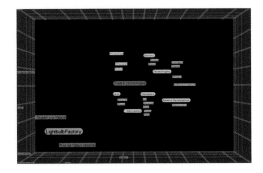

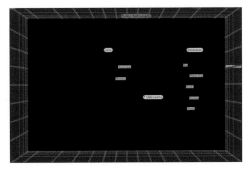

In 1996, researchers at Xerox Palo Alto Research Center developed a novel spatialization of the Web in order to provide users with a more effective tool to organize and browse online information. The system had two distinct components. The first was WebBook, a three-dimensional interactive book of HTML pages. WebBook allowed for rapid interaction with objects at a higher level of aggregation than pages. The second was Web Forager, an application that embeds the WebBook and other objects into a hierarchical 3-D workspace. The Xerox system applies very literal metaphors of familiar physical objects – the book, the desk and the bookcase – as a way of organizing large volumes of unordered Web pages.

In the conventional Web browser, the user sees only a single page at a time. With WebBook, large collections of pages can be browsed and quickly flicked through. Its creators argue: "We use the book metaphor not primarily because it is familiar, but because of the operational match to a corpus of interest and the efficient display characterization."

Web Forager was an experiment in what its creators called "task-tunable information space" and it had three distinct levels of "attention". The first level was for direct interaction, represented by the open book floating in a user's line of sight. The second level is given by the books on the desk and floating in the air – called "immediate memory" because users can quickly explore these information sources. The third level is the bookcase, where information can be stored in an orderly fashion for later retrieval. All objects can be moved in the three-dimensional environment, for example touching the bookcase automatically "flies" the user to directly in front of it, so that a book can easily be accessed. The researchers conclude:

The Web Forager workspace is intended to create patches from the Web where a high density of relevant pages can be combined with rapid access . . . Through the invention of such techniques and analytical methods to help us understand them, it is hoped that the connectivity of the Web, which has been so successful, can be evolved into yet more useful forms.

3.32: WebBook and Web Forager 3-D information spaces
chief cartographers: Stuart Card, George Robertson and William York (User Interface Research Group, Xerox Palo Alto Research Center, USA).
aim: to provide users with a more efficient means to organize and browse the Web.
form: literal metaphors of a book to hold a collection of individual Web pages. The books are then organized in a 3-D space with a desk and bookcase.
technique: 3-D interactive graphics and animation.
date: 1996.
further information: see <http://www.parc.xerox.com/istl/projects/uir/>
further reading: "The WebBook and the Web Forager: An Information Workspace for the World-Wide Web" by Stuart K. Card, George G. Robertson and William York, ACM Conference on Human Factors in Software (CHI '96), ACM, pp. 111–117. <http://www.acm.org/sigchi/chi96/proceedings/papers/Card/skc1txt.html>

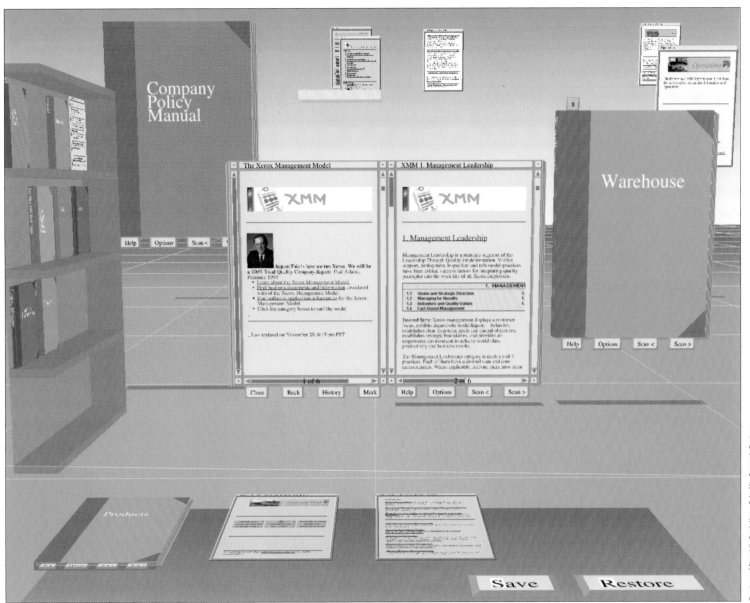

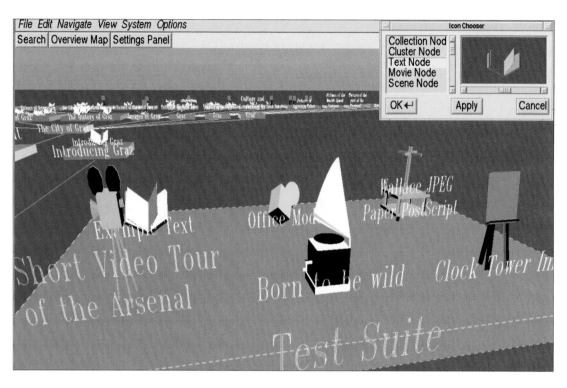

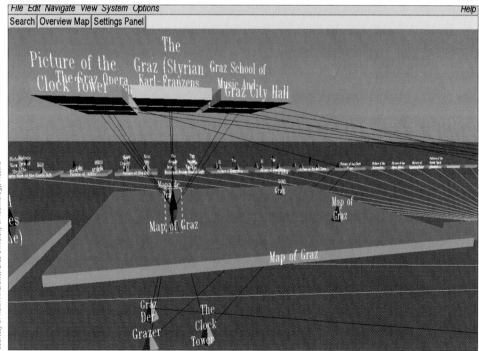

The images here show the Harmony browser that a client has used to access an Internet hypermedia system called Hyperwave. Harmony provides an integrated 2-D graph map and 3-D landscape view of the structure of a hypermedia information space in addition to the conventional page view of the documents. Here, we detail the 3-D information landscape component developed by Keith Andrews and colleagues at the Graz University of Technology in Austria.

3.33: Harmony information landscape

chief cartographer: Keith Andrews (Graz University of Technology, Austria).
aim: to provide a 3-D interface to hypermedia data.
form: 3-D landscape with iconic representations.
technique: converts a hierarchical structure into a landscape.
date: 1994–1995.
further information: Keith Andrew's homepage at <http://www2.iicm.edu/keith>
further reading: "Visualizing Cyberspace: Information Visualization in the Harmony Internet Browser", by Keith Andrews, in proceedings of IEEE Symposium on Information Visualization (InfoVis 95), October 1995, Atlanta, USA, pp. 97–104.
<ftp://ftp.iicm.edu/pub/papers/ivis95.pdf>

The Harmony Information Landscape provides a 3-D spatialization of an information space for users to browse resources (such as documents, files, images, etc.) which are represented by blocks and icons laid out across an infinite flat plain. Collections of resources (the Hyperwave equivalent of a website) are represented by flat slabs onto which the actual resources are placed as iconic glyphs, such as a book to represent a text file and an old-fashioned movie camera to identify a video clip (see top-left). The spatial arrangement of the blocks encodes the hierarchical structure of the Hyperwave information space. The user is able to fly over the landscape and choose objects of interest, which are then displayed in the conventional browser window. As the user browses, new collections are added dynamically to the landscape. In this way, the hyperwave site can be browsed and navigated as a 3-D landscape.

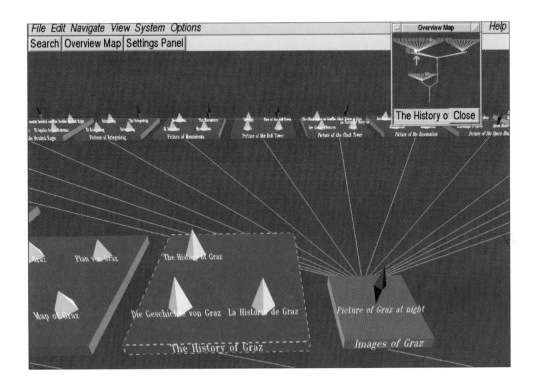

VR-VIBE represents the application of Collaborative Virtual Environment (CVE) technologies for information searching and retrieval, creating a 3-D cooperative system that can be simultaneously shared by several users. (Internet virtual worlds, such as AlphaWorld, are commercial examples of CVE, and we consider them in detail in chapter 4.) The screenshot opposite shows a pyramid of blocks, each representing a document in the bibliography. Matching documents from keyword queries are displayed as simple blocks floating in patterns above a flat landscape covered with a regular yellow grid. Keywords, represented as octahedra, are positioned across the space and the document blocks are displayed in relation to strength of attraction to each keyword. As such, keywords act as virtual magnets pulling documents toward them with differing strengths depending on their significance to the search. The images here display screenshots of a VR-VIBE session where over 1,500 documents are spatialized according to five keyword "magnets". The size and color of a document block encodes the relevance score of that document to the overall query, so large, brighter blocks that are visually the most prominent are the best matches to the whole query.

Crucially, this data space is a shared virtual environment. Other users present in the space are represented by simple shapes that look like sticks with eyes. This shared aspect raises interesting possibilities for collaborative searching and exploration of a large information space. Users are able to dynamically interact with the VR-VIBE data space in a number of ways. For example, a user could fly in close to examine part of the document space in detail and then quickly fly above to get an overall view of the configuration. Users can change the parameters of the search query by adding, deleting and moving the keyword "magnets". So the user can select a keyword glyph and move it by dragging it, and then the spatial arrangement of the documents will dynamically adjust. It is also possible to dynamically change the thresholds of the query using a 3-D scrollbar to limit the number of matching documents. Finally, documents of interest can be selected, and these can be fetched and displayed in a conventional Web browser.

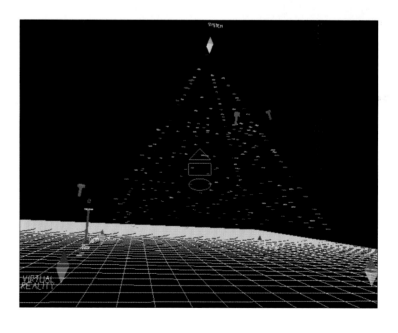

3.34: VR-VIBE, an example of a populated information terrain

chief cartographers: Steve Benford and colleagues (Communications Research Group at the University of Nottingham, UK).
aim: to visualize and interact with relational databases.
form: a 3-D landscape of floating blocks.
technique: calculates strength of attraction to keywords.
date: 1995.
further information: see <http://www.crg.cs.nott.ac.uk/>
further reading: "VR-VIBE: A Virtual Environment for Co-operative Information Retrieval" by Steve Benford, Dave Snowdon, Chris Greenhalgh, Rob Ingram, Ian Knox and Chris Brown, paper for Eurographics '95, 30 August – 1 September 1995, Maastricht, The Netherlands, pp. 349–360.

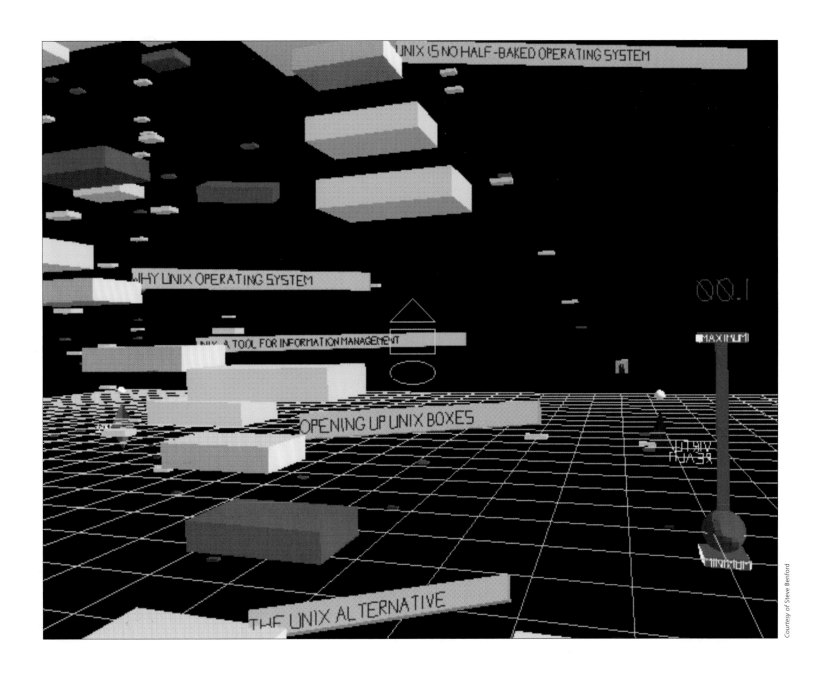

UNIX IS NO HALF-BAKED OPERATING SYSTEM

WHY UNIX OPERATING SYSTEM

A TOOL FOR INFORMATION MANAGEMENT

OPENING UP UNIX BOXES

THE UNIX ALTERNATIVE

00.1

MAXIMUM

MINIMUM

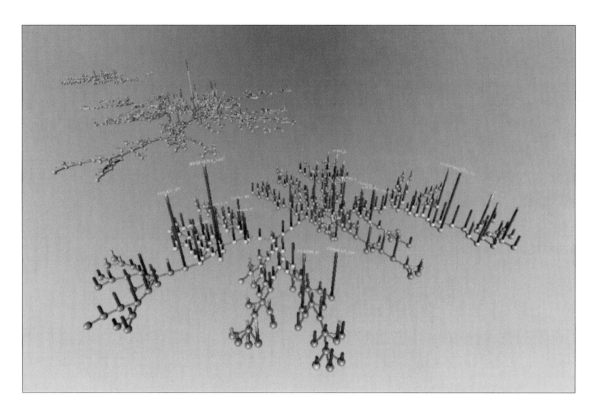

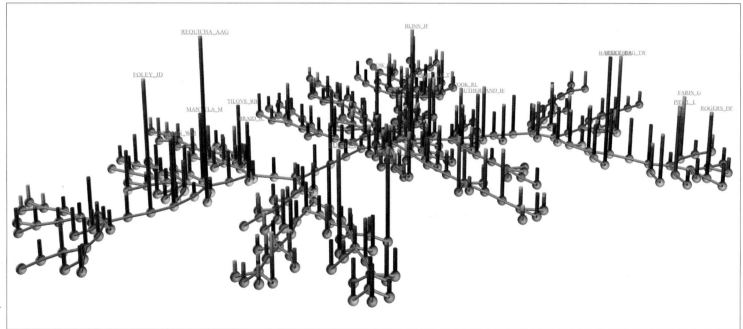

The images here show three sets of "semantic constellations" created by Chaomei Chen, in the Department of Information Systems and Computing at Brunel University, United Kingdom. The model is designed to help users browse through a complex collection of electronic documents, is created via VRML, and is used within a standard desktop VR environment. The images are example spatializations of an information space containing large collections of documents (academic conference proceedings). In the latter case, every paper in the collection is shown as a sphere in the semantic space, a metaphorical constellation of documents. The VRML display reveals how these papers are semantically connected in this semantic space.

As with HyperSpace, an arc–node 3-D graph is employed, with spheres representing the individual papers and the arcs connecting them based on how related their content is. The spheres are color-coded by year. Unlike HyperSpace, the arcs are not explicit "hard-coded" hyperlink connections; instead, in the constellation they are based on a computed measure of semantic similarity between the papers. So the papers that discuss the same or related topics will be semantically linked in the spatialization. The more closely two papers are related in terms of their content, the nearer they are in the semantic space. The semantic linkages are spatially arranged and connected into what is known as a "PathFinder" network. Pointing to a particular sphere causes the paper title to be displayed in a pop-up window and clicking on a sphere will display the paper abstract in a linked window in a Web browser. This allows an examination of the detail of a particular paper in one window and yet the ability to keep open the entire semantic space in another.

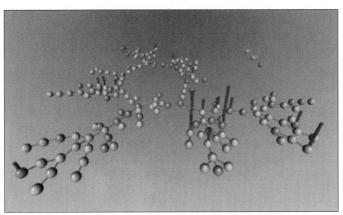

3.35: Domain visualization

chief cartographer: Chaomei Chen (Department of Information Systems and Computing at Brunel University, UK).
aim: to provide an interactive 3-D environment to explore large and complex collections of digital documents such as conference proceedings.
form: 3-D graph floating in space representing the document collection, with each sphere being a single document and the connections between them showing the semantic linkages. The vertical stacked bars projecting from the spheres show the volume of citations for successive time periods.
technique: sophisticated algorithms to analyze the semantic and citation. 3-D environment produced using VRML.
date: 1999.
further information: see <http://www.brunel.ac.uk/~cssrccc2/>
further reading: "Structuring and Visualizing the WWW with Generalized Similarity Analysis", by Chaomei Chen from Proceedings of the Eighth ACM Conference on Hypertext, June 1997, Southampton, UK, pp. 177–186.
<http://www.brunel.ac.uk/~cssrccc2/papers/ht97.pdf>

In 1995, Tim Bray spatialized large sections of the Web in order to answer four key questions, namely: How big it is? How wide is it? Where is the center? How interconnected is it? In order to answer these questions, Bray used a large search-engine index to calculate the key metrics on the structure of the known Web in 1995 (then a mere 11 million pages spread across 90,000 sites). These questions are still very relevant to academic researchers and commercial developers today. Much of the research into understanding Web morphology focusses on the analysis of the human-built hyperlink structures, the aim being to improve current Web searching tools and develop new searching algorithms to find elusive information sources that lie "hidden" in the ever-growing expanse of the Web.

Examining the hyperlink structures of the Web, Bray found that interlinking between sites was surprisingly sparse. Most links were local, within a site, and a few key sites (e.g., Yahoo!) acted as superconnectors tying sites together. He derived two intuitive measures of the character of a website, based on hyperlinks; these measures were "visibility" and "luminosity". Visibility is a measure of incoming hyperlinks – the number of external Web sites that have a link to a particular site. In 1995, the most visible website was that of the University of Illinois, Urbana-Champaign (UIUC), the home of the Mosaic browser. The vast majority of sites had very low visibility and nearly 5 percent had no incoming links at all. Measuring the reverse, the number of outgoing links, determines a site's luminosity. The most luminous sites carry a disproportionate amount of navigational workload. In 1995, Yahoo! was the most luminous site – and it probably still is today.

Bray used visibility and luminosity to map the key landmarks of the Web in 1995, highlighting the largest, most visible and most connected websites. The result is an information landscape dotted with 3-D models, which he termed "ziggurats" (the word for ancient stepped pyramidal temples). Each ziggurat visualized the degree of luminosity and visibility of a single site, along with the size of the site and its primary domain (government, education, commercial, etc.). The basic graphical properties of the ziggurat – its size, shape and color – were used to encode these four dimensions. The overall height represented visibility, the width of the pole represented the size of the site, in terms of number of pages, the size of the globe atop the ziggurat indicated the site's luminosity, and color-coding displayed the primary domain (green for university, blue for commercial, red for government agencies). The ziggurats were also labeled with the domain name of the site for identification. The spatial layout of the ziggurats across the plane were based on the strength of the hyperlink ties between them. The model is constructed in VRML and can be "flown around". The images opposite display fields of ziggurats at the very core of the Web in 1995. Further from this core there would be many thousands of other ziggurats, but most would be minuscule in relation to those at the heart. Bray developed many of these ideas for the Map.net visual search directory.

3.36: Web space as a landscape of ziggurats

chief cartographer: Tim Bray (whilst working at OpenText, USA).
aim: to visualize the structure of the whole Web.
form: ziggurat structures representing the characteristics of individual websites, arranged on a flat infinite plane.
technique: VRML-generated landscape.
date: 1995/6.
further information: see <http://www.textuality.com>
further reading: "Measuring the Web" by Tim Bray. Fifth International Conference on the World Wide Web, May 1996, Paris.
<http://www5conf.inria.fr/fich_html/papers/P9/Overview.html>

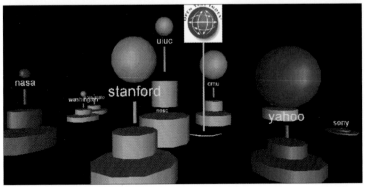

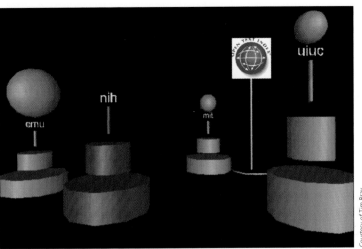

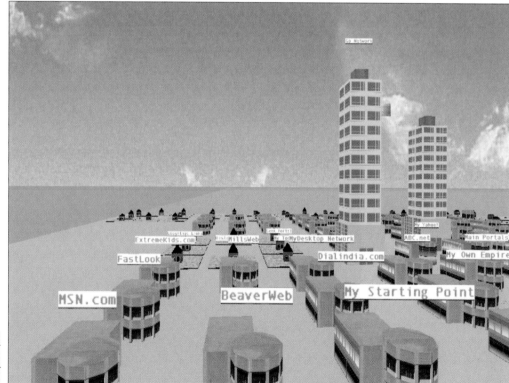

This spatialization is the three-dimensional equivalent of Map.net (detailed on page 122) and is a direct descendant of Bray's ziggurats (previous page). Instead of viewing an informational landscape as if from above, the 3-D version allows the user to walk across the landscape, navigating through buildings that represent particular websites. Although this restricts the field of view to a limited territory, each individual website can be scrutinized in more detail.

3.37: Visualizing Web space as a 3-D cityscape

chief cartographer: Tim Bray (Antarcti.ca).
aim: to provide a visual search engine.
form: 3-D cityscape, where each building represents an individual website. The shape and size of each building encodes characteristics such as the visibility of the website.
technique: 3-D interactive fly-through graphics.
date: launched November 2000. Screenshots taken in late November 2000.
further information: see <http://map.net>

Movement through the landscape is controlled by the cursor keys or the mouse, and the speed of movement can be altered; furthermore, it is possible to jump across the territory or teleport to new locations. Moving the mouse cursor over a building or area reveals an information box detailing the site name, a site description, and country of origin. The style, shape and size of the buildings encode information about the site's visibility and luminosity. Houses rank low on the visibility scale (top-opposite), office buildings are middle ranked (bottom-opposite), and skyscrapers have high visibility (below). Brown skyscrapers are designated "cool" by Open Directory editors. The size of the site is represented by roof size on a house, and height for offices and skyscrapers.

A very literal information landscape is the 3-D Trading Floor (3DTF) of the New York Stock Exchange, unveiled in March 1999. In 3DTF, information flows and real-time data are spatialized in a 3-D virtual environment modeled on a physical architectural space. Shown opposite are two striking views of the 3DTF environment.

The information environment is used as a real-time decision-support system for operators who manage the stock exchange, keeping vital networks and computer systems running as well as monitoring the actual information flows of the market performance. Users can be immersed in the interactive environment. Various real-time data streams on business systems – including stock performance of individual companies and user-defined aggregations – as well as not only the underlying networks and computer servers but also news broadcasts can all be spatialized. Particular failures, incidents and unusual activity can be highlighted and examined in detail.

3DTF is used by operators in a purpose-designed command center, nicknamed "The Ramp", in the heart of the actual stock exchange. It is displayed on a bank of large flat-screen panels. It is not accessible over the Internet. To power such a large and complex spatialization that is running with real-time data streams requires considerable computer resources, employing six expensive graphical supercomputers from SGI.

At present, 3DTF is somewhat of an experiment and it remains to be seen how valuable and useable the spatialization is in real-time management of the information space. There are plans to develop 3DTF further, with extended interaction and users represented by icons, as well as providing wider and remote access to the environment.

3.38: 3-D information environment – New York Stock Exchange trading floor

chief cartographers: Sabine Muller, Lise Anne Couture and Hani Rashid (Asymptote Architecture, New York).
aim: to manage the infrastructure and huge volumes of trading information for the New York Stock Exchange.
form: three-dimensional virtual environment with a range of graphical forms, charts and information panels and screen to spatialize a number of real-time data sources.
technique: environment written in VRML and powered by high-performance SGI graphics computers.
date: March 1999.
further information: see <http://www.asymptote.net>
further reading: "The NYSE's 3D Trading Floor" by Ben Delaney, IEEE Computer Graphics and Applications, November/December 1999, pp. 12–15.
<http://www.computer.org/cga/cg1999/pdf/g6012.pdf>

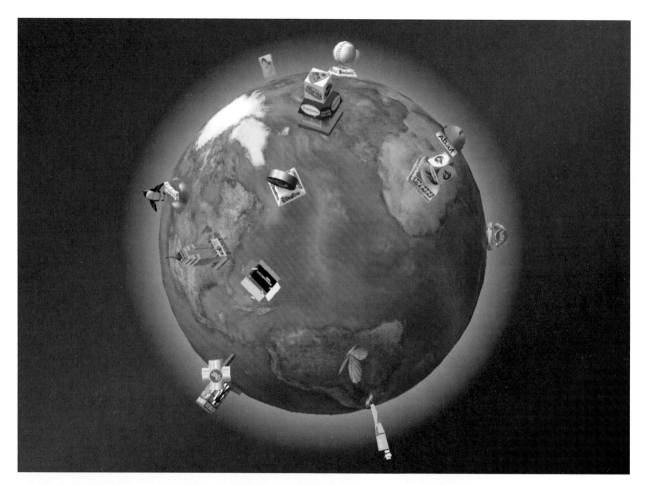

UBUBU seeks to provide a new desktop to Web browsers, using three-dimensional planets. The surfaces of the planets are populated with icons, such as skyscrapers and flower-pots, that provide visual bookmarks to Web pages, files and applications. UBUBU's planets are highly detailed and 3-D-rendered, floating in clear blue space. With this imagery, you can have the world as your desktop and the planets can be seamlessly rotated with the mouse. They are designed to make using a computer more fun by providing a more eye-catching desktop compared with more conventional interfaces.

The software comprises the (free) UBUBU Universe, which allows you to download a "solar system" of distinctive planets from a range freely available on the company's website. There are three classes of planets: topic planets, special-edition planets and community planets (submitted by users). Many of the planets and the default bookmark icons are sponsored by various corporations, such as SciFi channel, WWF wrestling, Warner Bros., etc. Shown here is the "Earth" topic planet (top-opposite) along with "Romance and Revenge" (bottom-left). The other planet (bottom-right) is a corporate special edition sponsored by the Hollywood film, *X-Men*, with the surface covered with photographs of the film's characters and the icons linking to related websites of the movie and cartoon.

Although the planets are customizable and users can delete the default icons and links, the whole effect has a strong marketing bias, with much of the effort being to get branding and advertising messages onto people's computer desktops. UBUBU earns revenue by users clicking on icons to go to a sponsor's website. It takes quite a bit of manual effort to reconfigure your own planet and override the corporate feel.

UBUBU's website claims that its software is "turning cyberspace into real space – and it's yours". However, the planets, whilst appearing to be effective spatializations, have little spatial consistency, with no relational quality between an icon's position and its neighbors. Indeed, on the default planets the icons are just scattered randomly. In this sense the planets are really just acting as eye-candy.

3.39: UBUBU

chief cartographer: Bryan Backus (UBUBU, pronounced "you be you be you").
aim: to make computers and the Web more fun by providing visual desktop with shortcuts and bookmarks.
form: 3-D planets covered with icons that provide bookmarks.
technique: custom-programed browser interface to the Web.
date: launched in April 2000.
further information: see <http://www.ububu.com/>

chapter 4

Mapping conversation and community

In this chapter we turn our attention away from examining attempts to spatialize information and network spaces and instead consider attempts to spatialize modes of online communication and interaction between people. As such, this chapter is concerned with what might be termed "people-centered" information visualization.

Cyberspace is composed of a variety of social media – email, mailing lists, listservers, bulletin boards, chat rooms, multi-user domains (MUDs), virtual worlds, game spaces – that support social interaction between people who are geographically dislocated. Many commentators now view the fostering of social interaction as the most significant aspect of cyberspace. Indeed, there is no denying that these social media are used every day and inhabited by millions of people talking, discussing, arguing, flirting, playing, and chatting with one another. As a consequence, the social uses and consequences of cyberspace have received significant academic and public attention, including attempts to spatialize these interactions in attempts to either engender more effective communication or to aid the comprehension of how these media function and the social relations they support. Before examining the wide range of ways to spatialize these media, we outline in brief some of the debates concerning the meaning and value of social media of cyberspace in order to illustrate its social significance and demonstrate the value of the spatializations we detail.

In general, analysts argue that the social media of cyberspace provide (1) new conditions under which individuals can explore and manipulate their identity; (2) new spaces in which communities – with very different characteristics to those in geographic space – can be developed and sustained.

In the first instance, analysts suggest that cyberspace allows individuals to explore their identity by changing the conditions under which identity is constructed and expressed. Cyberspace achieves this by providing a space of disembodiment and dislocation, because interaction is conducted through a medium that strips away body codings (e.g. age, gender, race) and geographic place and community. In other words, in cyberspace a person's identity is defined by words and actions, not body and place. In cyberspace, some commentators contend, your body is irrelevant and invisible and nobody need know your race, disability, gender, sexuality, material wealth, or geographic location unless you choose to reveal it. This stripping away, it is hypothesized, allows individuals to experiment with aspects of their identity that they would otherwise conceal. Although some question the degree to which cyberspace provides a space of meaningful social interaction, we believe that the influence of identity experimentation in cyberspace should not be dismissed lightly. The evidence gathered so far indicates that the social interactions that take place there clearly have a significant influence on some people, changing their outlook and values.

In the second instance, it is contended that individuals are exploiting the flexibility and fluidity of cyberspace to forge new communities – new social networks that are centered upon what they think, say, believe and are interested in. As such, some suggest that one of the principal effects of cyberspace is the formation of communities that are free of the constraints of place and are based upon new modes of interaction and new forms of social relationships. Instead of being founded on geographic propinquity, these communities are grounded in communicative practice. Here, individual participants can circumvent the geographical constraints of the material world and take a more proactive role in shaping their own community and their position within it, although (just like geographic communities) such online communities have behavioral norms, differing personalities, shared significance, and allegiances. Often, these communities are promoted as an antidote to, or as a supplementary means of belonging to or creating a sense of place for, traditional communities that are perceived to be disappearing in the geographic world due to processes of cultural and economic globalization, which are leading to a condition of placelessness.

In order to structure our discussion of the spatializations, we have divided the chapter into sections defined by the mode of social interaction (e.g., email, mailing lists, chat, MUDs, virtual worlds, games). These sections have been ordered in relation to whether they are asynchronous or synchronous modes of communication, and also taking into account their relative sophistication. Asynchronous media are domains in which interaction does not occur in real time (e.g., email, mailing lists, bulletin boards, Usenet news), and synchronous are the converse (e.g., chat rooms, MUDs, virtual worlds, multiplayer games). Outside cyberspace, letter writing is the archetypal asynchronous mode of social interaction and face-to-face communication is the archetypal synchronous mode. Many of the asynchronous media provide "persistent conversations" that continue to exist beyond their immediate posting.

Mapping email

Email was the first social medium of cyberspace, developed by Ray Tomlinson at BBN in 1971. It allowed users to post mail messages across a network to individual accounts. The supporting software was quickly circulated between all ARPANET sites and by 1972 it was one of the two most widely used applications on the network (the other being remote log-in services). Since then, email has remained the most popular, well-used and powerful of all social media.

A report by Peter Lyman and Hal Varian, both of the University of California at Berkeley, stated that over 610 billion emails were sent in 1999. Email can be read and sent from a networked computer, even over the slowest modem links. Its standing is such that, for most Internet users, the first thing they do when they log on is to check their email, and it has been credited in part with fueling the surge in Internet user growth as many people seek out an Internet account because of the lure of email communication with friends, family and colleagues. A recent survey from the University of California, Los Angeles found that 82 percent of Internet users use email. Of those, 76 percent check their email at least once a day.

Despite email's enduring popularity as a medium, there have been few attempts to spatialize the structure and content of email. Interfaces for email clients are still pretty much the same as a decade ago. As befits a predominately text-based form of communication, messages are stored in sorted lists and arranged in folders. And yet there is increasing need for better tools to manage the inexorable growth in the volume and importance of the email that many people receive on a daily basis. Two interesting prototype email clients that take a much more visual approach to reading, composing and managing messages are Parasite and ContactMap, described next.

One interesting experiment in designing a visual email application was Parasite. The visual metaphor used to map out the storage of messages in mailboxes is an adaptation of the Feynman diagrams normally used in the field of particle physics. The interface looks complex, with a multilayered display, and it focusses on revealing the connections between messages and the people who sent them. Cannon and Szeto explain:

As each participant contributes to a given topic over time, links thread an email "conversation" together. By emphasizing these links, Parasite preserves the relational quality of each bit of text so, as an online correspondence builds up, the interface gradually maps out a "community memory" of the discussion.

Unfortunately, development of Parasite ceased before a finished version of the software was released, and so it will remain an interesting but untested approach.

ContactMap is development software that allows users to manage and examine their communication. It extends beyond an email client to allow click-to-dial instant-messaging video and fax. It also facilitates finding documents by contacts, taking memos, and monitoring the availability of contacts.

The software can also be used to examine the interactions between people, using the information to construct a map of contacts, and to perform an analysis of the contacts and their importance. In order to create an "importance rank" it weights factors such as "number of replies", "number of sent messages", "who replies to your email", "who sends you messages", "who is often mentioned".

The spatialization opposite shows the social network of Bonnie Nardi, one of the software's developers, with her contacts grouped by location. It is essentially a visual address book.

4.1: Parasite

chief cartographers: Steve Cannon and Gong Szeto (whilst at i/o360).
aim: to provide a visual email client.
form: a complex, multilayered graph where blue dots represent individual messages and wavy lines show connections.
technique: adaptation of Feynman diagrams.
date: 1998.
further information: see <http://www.ure.org/>

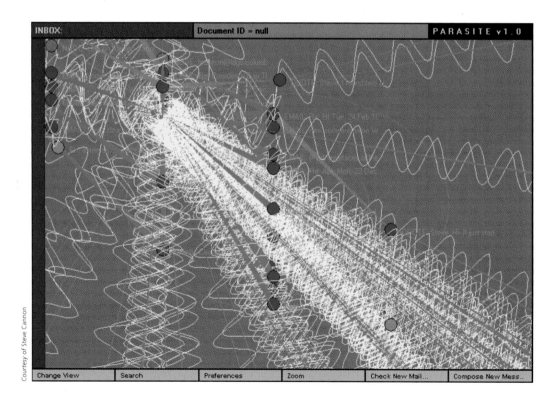

Courtesy of Steve Cannon

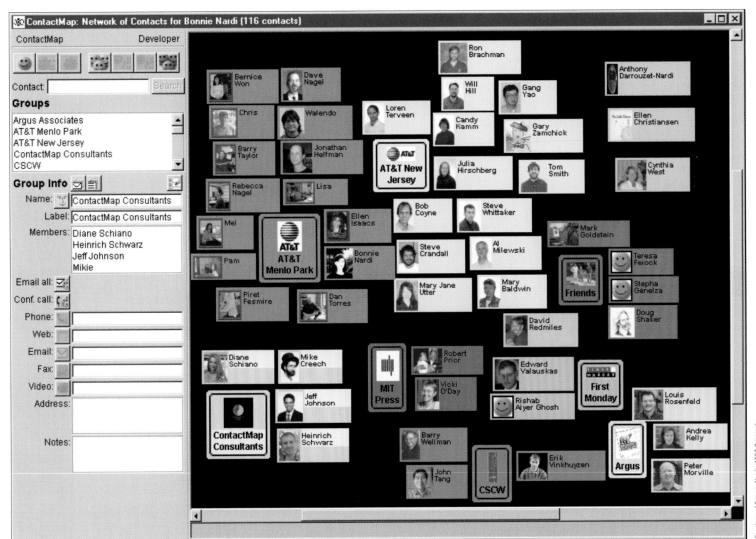

Courtesy of Bonnie Nardi, AT&T Research

4.2: ContactMap

chief cartographers: Bonnie Nardi, Steve Whittaker and Ellen Isaacs (AT&T Research Labs).

aim: to create a social desktop that integrates a person's information and communication interactions.

form: a kind of simulated desktop where different people are shown by small iconic business cards.

technique: custom-written browser and contact-manager software.

date: 2000.

further information: see <http://www.research.att.com/~stevew/>

Mapping mailing lists and bulletin boards

The power of email for one-to-one communication can easily be used for one-to-many interactions and many-to-many discussions. This is achieved through the use of mailing lists, listservers and bulletin boards. A mailing list is a one-to-many communication medium where the list owner can send messages to all members of a list. A single message can therefore be delivered to hundreds of subscribers with no extra effort or costs. Examples include commercial daily news bulletins (e.g. Wired News at <http://www.wired.com>) or weekly sarcastic Net gossip (e.g. Need to Know! at <http://www.ntk.net>). A listserver extends this idea to allow many-to-many discussions by permitting all subscribers to post messages to the list. This opens the way for ongoing conversations involving many participants. Bulletin boards, similarly, allow many-to-many communication between individuals. However, unlike mailing lists, messages are not redistributed to subscribers; instead, messages are posted to a central site, now usually Web-based, which users have to log on to in order to check messages.

Many tens of thousands of different mailing lists, listservers and bulletin boards exist, covering almost every imaginable topic. As with email, there have been relatively few attempts to spatialize them, but we detail next three interesting examples: Visual Who, PeopleGarden and WebFan.

The mailing lists that people subscribe to can reveal much about their interests and the form of the wider online social communities in which they participate. Visual Who is an interactive mapping tool designed to dynamically spatialize the social patterns (based on affinities) of a large online community, as measured in mailing-list usage.

The pictures opposite show screenshots of the Visual Who mapping system displaying the affinity of a large number of people to different mailing lists, such as softball, agents and holography. The lists are positioned as anchor points around the edge of the spatialization window, and the people are represented by their names. The position of people's names is relative to the strength of their affinity to each of the anchor mailing lists: people with a strong affinity to a certain list will be drawn close to it on the spatialization, whereas someone with approximately equal affinity to two lists would be drawn to a midway point between each. As a case study, Judith Donath used the 700 or so people affiliated to the MIT Media Lab. The color-coding of people's names is based on their work status (faculty are yellow, staff are purple, graduate students are red, and undergraduates are green). Visual Who is also an interactive spatialization in that the user of the system can move anchors, delete and add new lists, with the mapping being rearranged dynamically to take account of the new forces. The system can also show presence by only mapping the people who are actively logged on to the computer system at any one time. In this way, a user can explore quantitatively the social patterns and uses of list space.

We look at more of the visualization work of Donath's research students over the next few pages (PeopleGarden, WebFan, Loom, and Chat Circles).

4.3: Visual Who

chief cartographers: Judith Donath, with Dana Spiegel, Danah Boyd and Jonathan Goler (Sociable Media Group, Media Lab, MIT).
aim: to show the complex patterns of social connections between large numbers of people, based on their mailing list affiliations.
form: people are represented by their names, which float in a black 2-D space.
technique: spatial position of people's names is determined by their attraction to the mailing-list anchors, based on the strength of their affinities.
date: original version 1995.
further information: see <http://smg.media.mit.edu/projects/VisualWho/>
further reading: "Visual Who: Animating the affinities and activities of an electronic community", by Judith Donath, in Proceedings of ACM Multimedia '95, 5–9 November 1995, San Francisco. <http://smg.media.mit.edu/people/Judith/VisualWho/>

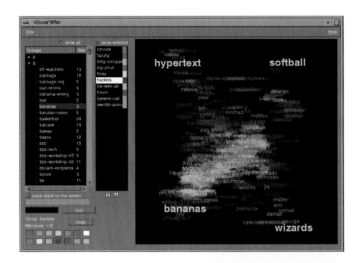

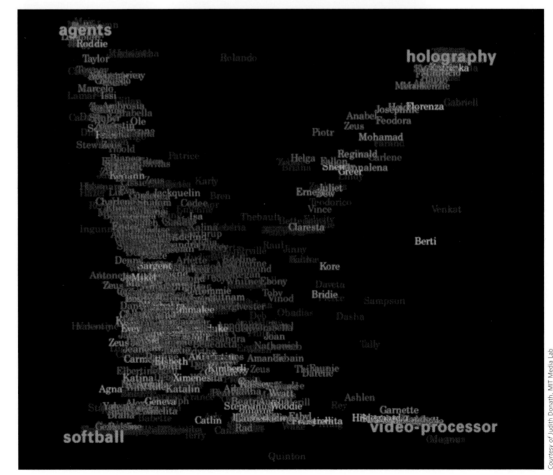

Mapping conversation and community

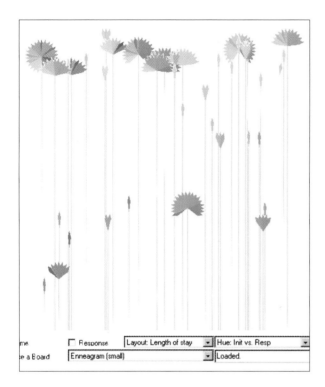

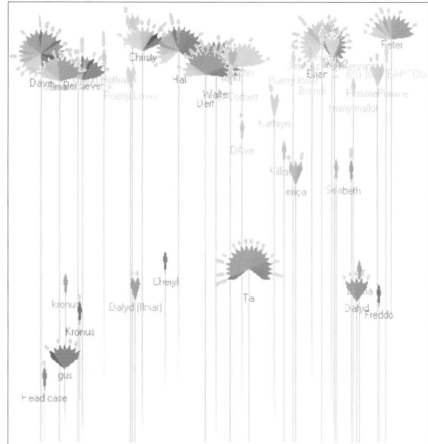

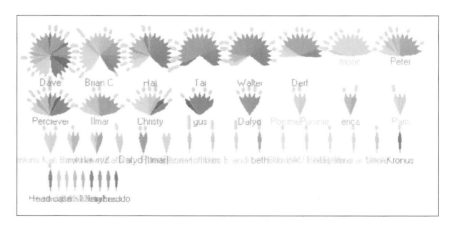

Rebecca Xiong has developed a number of innovative interactive visualizations of social interactions in cyberspace (including Netscan cross-visualization). The problem of how to gain a sense of the social nature of discussion spaces is difficult, particularly with conventional interfaces that just present a hierarchical listing of text messages. Xiong has sought to tackle this problem by spatializing the patterns of postings. This, she suggests, would enable users to answer basic questions such as "How much interaction is there?", "Who are the 'experts' and 'old-timers'?" "Is the space friendly for newcomers?" and "What is the style of conversation?"

At the heart of Xiong's spatialization work is the use of what she terms "data portraits" to represent people in the discussion space. She explains: "Unlike photo-realistic portraits, which show physical features such as gender, age or race, data portraits are abstract representations of users' interaction history." They are not fixed but change over time as the person's social relations evolve. By displaying individual data portraits of discussion members on a single screen, it is possible to make quick visual comparisons of people and, most importantly, gain a sense of the overall "shape" of the community at a particular snapshot in time. PeopleGarden and WebFan both display Xiong's data portraits.

4.4: PeopleGarden and WebFan

chief cartographer: Rebecca Xiong (whilst a graduate student at Lab for Computer Science, MIT).
aim: to spatialize Web-based discussion boards in order to more clearly see patterns of social interaction over time.
form: PeopleGarden uses visual metaphors of a flower to represent individuals and collections of flowers in a garden to show the whole discussion space. WebFan shows messages as short line segments arranged in a branching fan-shaped layout.
technique: interactive graphics using Java.
date: PeopleGarden – 1999; WebFan – 1998. Screenshots taken October 2000.
further information: see Xiong's homepage at <http://pigment.lcs.mit.edu:8080/~becca/>
further reading: "PeopleGarden: Creating Data Portraits for Users", by Rebecca Xiong and Judith Donath, ACM Symposium on User Interface Software and Technology 1999, pp. 37–44. Also at <http://pigment.lcs.mit.edu:8080/~becca/papers/pgarden/uist99.pdf>

In PeopleGarden (opposite), the data portraits of individual discussion participants are called PeopleFlowers. Each petal in a flower represents a single message. Xiong says that she chose the flower metaphor because she liked "the organic nature of a flower, and the suggestion that it changes over time, as users do". Indeed, it provides a simple, yet attractive, visual metaphor. The form and color of the flower encode data on the number of messages posted, their temporal sequence, whether they were initial conversations or replies, and also the number of replies garnered in return from other participants.

As messages are posted, more petals appear and the flower blooms into life. The more active a user is, the more significant is the flower that is "grown". The petals are positioned clockwise with age and the color of each petal fades with time, so that the newer postings are indicated by brighter petals. Also, a distinct change in the color saturation of the petals indicates a significant time gap between postings. To show the number of replies a message receives, small yellow dots are added to the tip of the petal. Finally, strong color-coding can be enabled to distinguish petals that represent initial postings and those that are replies. In the images opposite, the magenta-colored petals are new initial messages, while the blue ones are replies.

Displaying all the individual PeopleFlowers of a discussion space creates what Xiong calls a PeopleGarden. She argues: "We have used the garden metaphor because a healthy garden has certain properties that we can use to represent a healthy discussion group." For example, if a PeopleGarden contains only a few small flowers, with sparse and faded petals, one could conclude that the discussion space is relatively lifeless. The height of each flower in the PeopleGarden represents the length of time that a particular user has been involved (since a first posting). Thus, the tallest flowers are the oldest, fitting well with the overall organic metaphor. The patterning of flowers at different heights indicates how well the space is attracting new users. The horizontal arrangement of flowers is random. One can also easily compare flowers, and thereby people's interactions – seeing who is most active, who replies a lot, and who receives most replies to their messages. This comparison can be facilitated by choosing the option to lay out flowers according to the total number of postings. There are also several other useful options – for example, one can toggle the display of user names below the flowers, the display of the yellow reply markers, and choose between different layouts.

An earlier, somewhat simpler, visualization developed by Xiong is called WebFan, where a semicircular fan of small lines represents messages posted on a Web-based discussion board (below and opposite). One of the principal aims of WebFan is to show the dynamic nature of discussion over time, by representing the detail of posting and (especially) reading messages. Animation is used to show the reading of messages by different users throughout the day, on top of the fan structure. The initial message in a thread is the line segment at the very center of the semicircle; subsequent replies are added as branches to the end of the line; new messages are added at the top of the semicircle. The fan of message segments is interactive, so that moving the mouse cursor over a line causes the area immediately around it to expand, ensuring greater visibility, with the title, author and date of the message simultaneously displayed in a pop-up box. Right-clicking on the message segment causes the actual text of the message to be displayed in another frame of the browser.

Enabling the animation of WebFan shows social activity by color-coding the line segments based on who has read them. (Each participant has a unique color.) If a message is read by several people, it becomes multicolored. As time progresses it is possible to see individuals move through the fan (represented by a small circle symbol) and the line segments change color as they are read. Also, on the right-hand side of the fan a simple activity chart can be scrolled down. The differently colored bars represent people, and the length of the bar is the amount of time that a particular person has spent reading/posting messages.

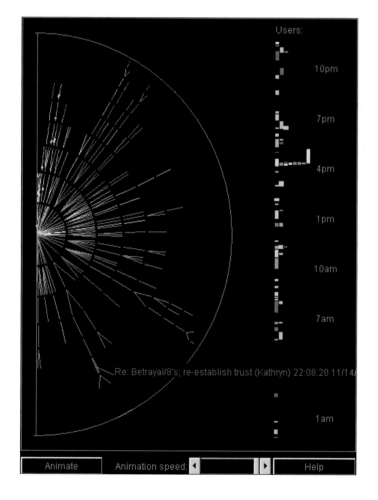

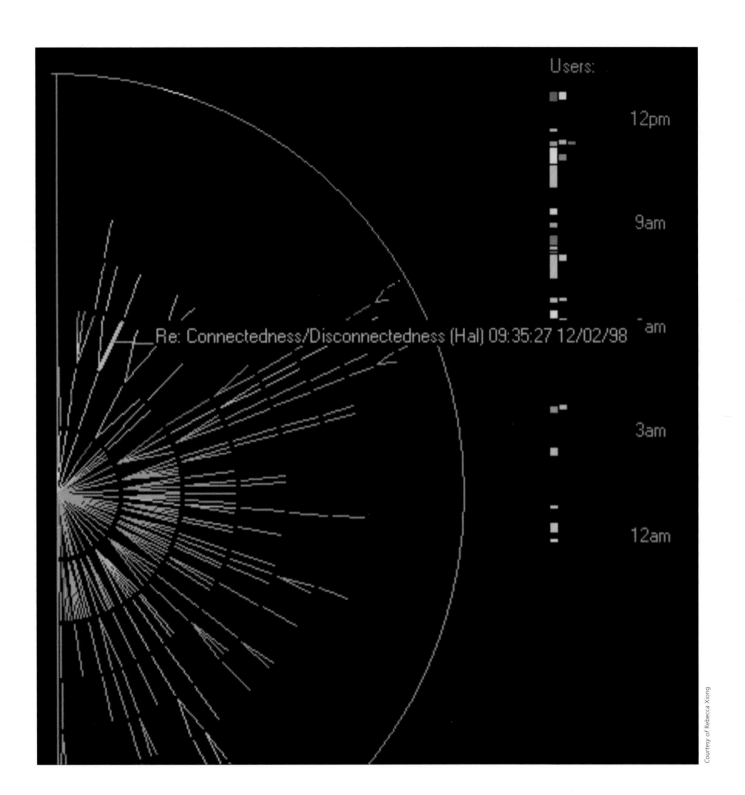

Re: Connectedness/Disconnectedness (Hal) 09:35:27 12/02/98

Users:

12pm

9am

am

3am

12am

Mapping Usenet

Usenet is the most popular discussion space on the Internet. It was developed in 1979 by Jim Ellis and Tom Truscott at Duke University and Steve Bellovin at the University of North Carolina. It is a vast, distributed bulletin board running on top of the Internet, and it provides a complex mesh of interrelated spaces known as newsgroups. (See page 53 for old maps of Usenet traffic flows.)

There are many thousands of different newsgroups, covering a huge range of topics. Newsgroups are arranged in a number of large hierarchies based on very broad classifications. The seven of the oldest hierarchies are *comp, sci, soc, talk, news, misc, rec*, later joined by the largest and most contentious of all – the *alt* (alternative) hierarchy. The *comp* hierarchy contains the newsgroups for computer-related discussion, *rec* is for hobbies and recreational activities, while *alt* holds a miscellaneous bunch of groups (including what some see as the controversial groups like *alt.drugs* and *alt.sex*). There are also hierarchies for major organizations, commercial software vendors, countries, and local site-specific groupings. Each newsgroup holds many separate, ongoing conversations – usually with multiple participants.

We look at two prototype newsreaders, Loom and Conversation Map, which try to spatialize messages to enhance users' understanding of the wider social context of a newsgroup. We then examine Netscan, an ambitious five-year project led by sociologist Marc Smith to provide a comprehensive array of Web-based tools to analyze the social geography of Usenet.

Loom aims to visualize the patterns of article posting within individual newsgroups over time, with a focus on individual participants and the article threads. A variety of spatializations, all using a simple two-dimensional grid structure, can be produced to show the different social structures underlying posting interaction in the newsgroup, as displayed opposite. Within each image shown, the horizontal axis represents time and the vertical axis is divided into individual columns, each representing a single group participant.

Top-right, each message is represented by a single colored symbol in the spatialization – the color can be used to represent an important characteristic of the article, such as subject, or the domain of the poster. The lines connect messages of a single thread. At first glance, the spatialization is difficult to read; however, according to the developers of Loom, there are three key patterns to look for that aid interpretation: first, a strong vertical patterning of symbols represents an intensity of activity at a particular time; second, prominent horizontal rows of symbols show the most active participants; third, the structure of connecting lines between articles is an indication of the conversational atmosphere of the group, for where there are long, complex and overlapping lines one can infer a group of intense discussions involving many participants with many replies and long threads, and where the lines are much shorter and form a more disconnected mesh, this indicates a more "question and answer" style of interaction.

The second image (bottom-right) shows the Loom spatialization concentrating on the temporal patterns revealed by Loom. The solid vertical lines are user-specified time units (such as days or weeks) and the dots at different heights are the articles posted by different people.

The last image (left) is an attempt to classify and spatialize the actual content of the articles rather than just the structure of the group. Automatic content classification is a major challenge facing those wishing to map social cyberspace. Loom uses a heuristic device to classify articles into four categories: angry, which is mapped as red, peaceful (green), news-based (yellow), and all others (blue). The spatialization is divided into a cellular structure based on the boxes of a calendar. Each day's articles are represented by appropriately colored disks in the cell. Clicking on a particular disk will cause the text of an article to be displayed in a pop-up window. The spatialization displays a whole month's worth of newsgroup posting and it is possible to see the daily intensity and tone of conversations from the density and color of the disks. In this way, this particular type of Loom spatialization acts as a kind of visual overview, and also generates a unique signature of the group.

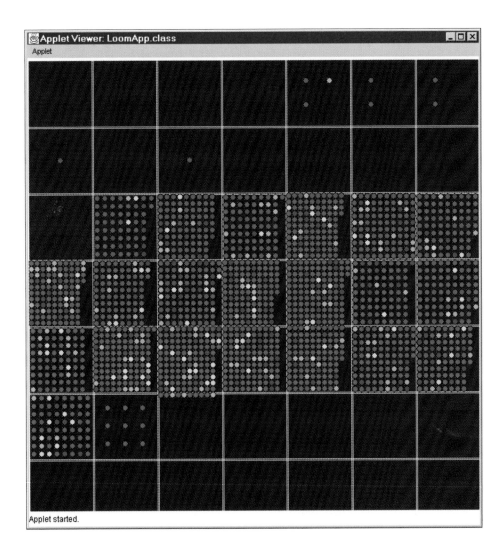

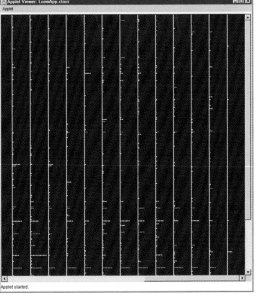

4.5: Loom

chief cartographers: Karrie Karahalios, Judith Donath and Todd Kamin (Sociable Media Group, Media Lab, MIT).

aim: to provide a newsreader that is able to visualize the patterns of postings in a newsgroup.

form: structured 2-D information maps, where one dimension is people or messages and the other dimension is time.

technique: custom-written software to analyze messages by poster and time. Spatializes with interactive graphics using Java.

date: 1998.

further information: Sociable Media Group homepage at <http://smg.media.mit.edu>

further reading: "Visualizing Conversation" by Judith Donath, Karrie Karahalios and Fernanda Viégas, *Journal of Computer-Mediated Communications,* Vol. 4(4), June 1999. <http://www.ascusc.org/jcmc/vol4/issue4/donath.html>

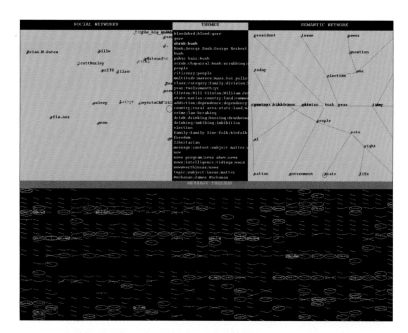

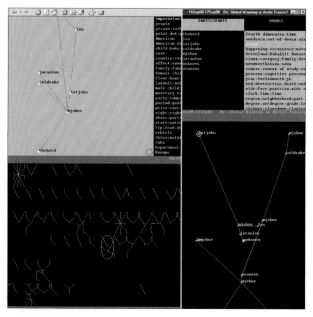

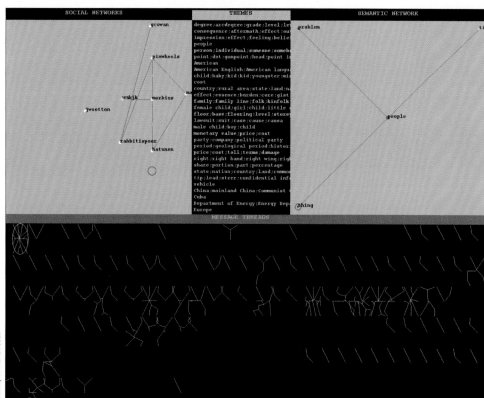

Conversation Map is a visual browser for large online discussion, such as Usenet. It is a prototype being developed by Warren Sack, initially as a doctoral research project at the MIT Media Lab. Sack is interested in analyzing and mapping the structure of the actual messages in what he terms "very large-scale conversations" (VLSC), where hundreds of participants may exchange thousands of message over a short time period. He says the spatialization provided by Conversation Map enables a user to "browse a set of Usenet articles according to who is 'talking' to whom, what they are 'talking' about, and the central terms and possible emergent metaphors of the conversation". It is therefore a browser of content that harnesses the analytical power of computational linguistic and interactive graphics to try to provide users with a richer and deeper view of social interaction than conventional Usenet interfaces.

The screenshots of Conversation Map spatialize the *sci.environment* and *alt.politics.elections* newsgroups. Initially, each interface looks somewhat complex, comprising four separate panels. These panels show different views of the same newsgroup, with the top half comprising graphical interfaces to the three key analytical tools – the social network, the discussion themes, and the semantic network. The bottom half of the screen shows all the message threads as small graphs. It is important to note that these four panels are interactive and fully interlinked, and so selecting a data element of interest in one panel will also highlight the same feature in the complementary views.

Conversation Map automatically calculates the social network of people in the newsgroup by analyzing the reciprocal exchanges of messages. This is visualized as an interactive spider graph, where individual participants are the nodes and their reciprocal connections are the lines. The participant's name can be toggled on and off for clarity. A user can explore the graph by clicking and dragging nodes with the mouse. The middle panel in each image gives a list of the key themes of the newsgroup, in descending order according to how often they occur in message threads. Again, this metric is generated automatically using standard linguistic techniques. The right-hand panel is the semantic network showing which significant terms are being used in the messages and how these are connected.

Message threads for the given time period are displayed in the bottom half of the interface. They are represented by small spider graphs, which are arranged in chronological order from the upper-left corner of the panel. The complexity of the spider graphs indicates visually the number of messages in that thread. Moving the mouse over a graph displays the date and subject in a pop-up panel, while clicking on a thread graph will highlight the participants in the social network panel and the terms used in the semantic network. Double-clicking causes a new window to pop up displaying the thread in greater detail, enabling one to explore further and actually read messages. In the example screenshot from *sci.environment* shown top-right, the user has clicked on the thread entitled "Global Warming or Arctic Freeze?".

Conversation Map is a powerful prototype tool, although it is perhaps too complex at present for average Usenet participants. As such, it may be more useful as an exploratory tool for sociologists and anthropologists who are analyzing online discourse.

4.6: Conversation Map

chief cartographer: Warren Sack (whilst a graduate student at the Media Lab, MIT).
aim: to provide a visual Usenet browser that offers analysis of the social and linguistic structure of a newsgroup.
form: simple spider graphs represent the social network of posters and the semantic network of the message for a given newsgroup. Four interlinked panels show different views of the social content of the newsgroup.
technique: applies linguistic analysis to derive social and semantic networks. Visualization interface uses Java.
date: 1999. Screenshots taken in November 2000.
further information: see <http://www.sims.berkeley.edu/~sack/>
further reading: "Conversation Map: A Content-Based Usenet Newsgroup Browser", by Warren Sack, in Proceedings of the International Conference on Intelligent User Interfaces, ACM, January 2000, New Orleans. <http://www.media.mit.edu/~lieber/IUI/Sack/Sack.html>

One of the leading researchers analyzing and visualizing social interaction in cyberspace is Marc Smith, a research sociologist at Microsoft Research. His ongoing project is called Netscan, and it involves measuring and mapping the social structures of Usenet, at various scales ranging from individual message threads, to newsgroups' interconnections, to maps of whole sections of Usenet space. The project began as part of Smith's doctoral research at University of California Los Angeles in 1994. This ambitious project monitors "every message from every newsgroup every day", equivalent to some 27,000 newsgroups carrying around 675,000 messages per day. Netscan therefore represents one of the most comprehensive projects analyzing and mapping social cyberspace.

At a basic level, Netscan data-mines daily Usenet traffic to produce a wealth of descriptive statistics on newsgroups and postings. This data is freely available on the Netscan website in summary tables and newsgroup scorecards (as shown bottom-opposite). Netscan also provides three more advanced analysis and visualization tools, these being Dashboard, the Cross Post network visualization, and treemaps of Usenet space.

Netscan Dashboard was released as a prototype application in Fall 2000 to provide users with a detailed and interactive analysis tool for examining the social communication that takes place in message threads of selected newsgroups. It was developed by Marc Smith and Andrew Fiore (a student intern). Dashboard is composed of several distinct visual interfaces, as can be seen in the screenshot top-opposite. The image shows the whole interface, comprising a selection panel, a tree visualization, a so-called "piano-roll" display, and an interpersonal connection graph. In this case the user is analysing a large thread from the newsgroup *microsoft.public.win96.gen_discussion*. The thread subject is "Bill Gates" and it comprises some 74 messages, posted over a four-day period by 26 different people.

One starts using Dashboard from the selection panel to isolate a particular message thread one wishes to examine. Different views of this thread are then displayed in the three panels. The main view is a tree visualization of the temporal sequence of the thread; each box represents an individual message and a line between boxes indicates that it was a reply to a previous one. The start of the message thread is the first box at the top of the tree, time moving forward as a user descends the tree. The gray banding represents separate days, and the width of the band is proportional to the number of messages posted on that day. Half-shaded boxes indicate that the message was from the most prolific contributors. Passing the mouse over a message box highlights the author in the other two panels. Clicking on a box causes the actual message to be displayed in a window at the bottom of the screen.

The piano-roll panel shows author activity over time. Each column shows a histogram of the number of messages that each author has contributed to the thread on each day. Again, the piano roll is interactive and interlinked to the other two panels. Clicking on an author provides access to an email facility enabling the user to contact that author directly. The final panel is a graph of interpersonal connections, where authors are represented as circles. The position of an author in the two-dimensional graph is based on the measure of the number of replies sent by the author and the number of replies received from other list members. Clicking on an author in the graph highlights, in the tree visualization, the messages sent.

4.7: Netscan Dashboard

chief cartographers: Marc A. Smith and Andrew T. Fiore (Collaboration and Multimedia Group, Microsoft Research).
aim: to visualize the social structures of message threads in a newsgroup. Part of the whole Netscan project for measuring and mapping the social geography of Usenet.
form: a three-panel presentation of different dimensions of the data.
technique: analysis of message threads over time, and the social network of posters. Spatialization by a Web-based interface using Java.
date: prototype launched October 2000. Screenshots taken November 2000.
further information: see <http://netscan.research.microsoft.com/>
further reading: "Visualization Components for Persistent Conversations", by Marc A. Smith and Andrew T. Fiore, MSR Tech Report 2000-98, September 2000.
<http://www.research.microsoft.com/research/coet/Communities/TRs/00-98.pdf>
"Invisible crowds in cyberspace: mapping the social structure of the Usenet" by Marc Smith, in Marc Smith and Peter Kollock (eds) *Communities in Cyberspace* (Routledge, London, 1999), pp. 195–219.

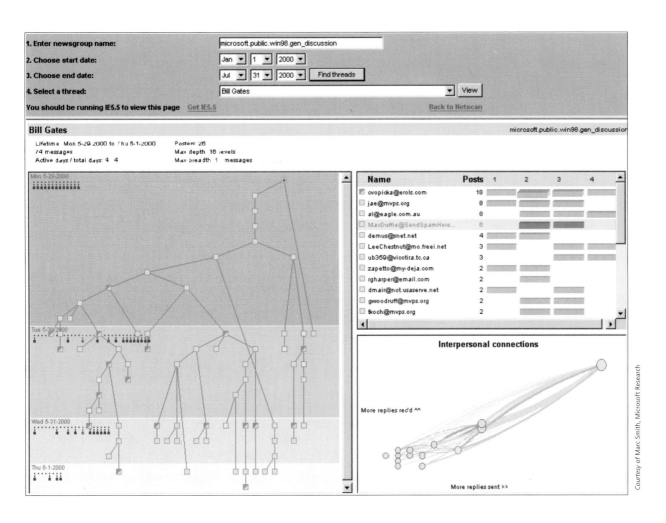

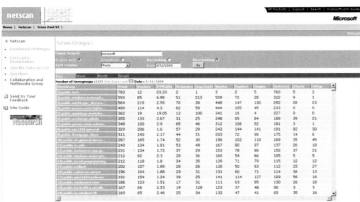

Courtesy of Marc Smith, Microsoft Research

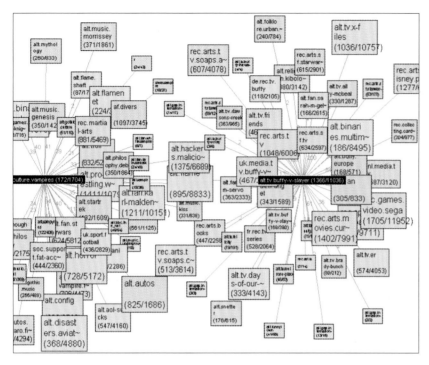
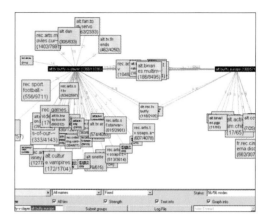
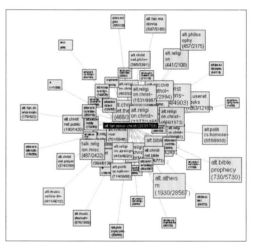
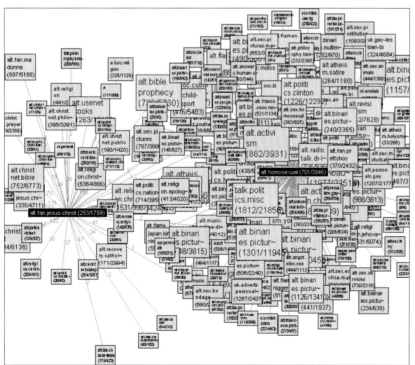
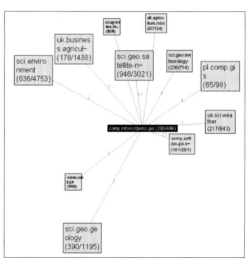

Courtesy of Marc Smith, Microsoft Research

Another part of the Netscan project is Cross Post, a network visualization that enables users to explore the connections between Usenet newsgroups. The system enables a user to view a cross-post graph for any of the thousands of newsgroups covered by the Netscan system, quickly revealing hidden clusters of related newsgroups. The visualization is also interactive, enabling a user to explore the graph and alter various settings.

The size and color of the boxes in the graphical representation (see maps on page 170 and below) represent the newsgroup's characteristics, with the size of the box proportional to the number of posts and the color showing the posts-to-poster ratio. The graph is also fluid, so that the user can move and position the boxes as desired. The graph layout is created using a simple "spring-based" algorithm, where boxes repel each other with the cross-post links acting as attractors and so working to counteract this. The lines between boxes show the ties between groups and can be labeled to indicate the strength of the relationship in terms of the number of cross-posted articles. The visualization of cross-posting is important because this is a key structural component of Usenet, which (Smith argues) forges small neighborhoods of interconnected groups. He found that only 6 percent of groups analyzed by Netscan stood completely alone, with no cross-posting at all. On average a newsgroup has some measure of connectivity to 50 other groups.

The image top-left shows the cross-post graphs for two potentially related groups: *alt.culture.vampires* and *alt.tv.buffy-v-slayer*. While both newsgroups have lots of connected groups, there appear to be few direct cross-post links. The map top-right shows two closely interlinked groups, namely *alt.buffy.europe* and *alt.tv.buffy-v-slayer*, which have many groups in common. Also shown on this image is the control panel that users can manipulate and tune the display. The groups connected to the newsgroup *alt.fan.jesus-christ* are shown middle-right, while the image in bottom-left shows the same group compared with *alt.homosexual*. The image bottom-right shows *comp.infosystems.gis*, a technical Q&A-orientated group for discussing geographical information systems (GISs). It has connection to 11 others, with the majority located in the *sci* hierarchy and concerned with physical environment applications of GISs.

The final example, below left, shows how the Cross Post visualization tool can be used to get a wider overview of a larger part of Usenet space by focussing purely on the patterns produced by the cross-postings of several large newsgroups. By turning off the display of names, one can reveal the structure of spaces around the group – what Marc Smith has tentatively termed the "electron shells". This scale of visualization, where one can see a large portion of Usenet, has been further explored using treemaps, described next.

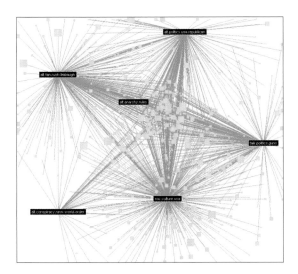

4.8: Netscan Cross Post network visualization

chief cartographers: Marc A. Smith and Rebecca Xiong (Collaboration and Multimedia Group, Microsoft Research).
aim: to provide an interactive tool that allows users to explore how newsgroups are connected by cross-posts. Part of the whole Netscan project for measuring and mapping the social geography of Usenet.
form: graph with colored, labeled boxes, each representing a single newsgroup. The lines show the cross-post links.
technique: interactive graphics using spring-based layout algorithms.
date: 1998. Screenshots taken November 2000.
further information: see <http://netscan.research.microsoft.com/>
further reading: "Visualizations of Collaborative Information for End-Users" by Rebecca Xiong, Marc A. Smith and Steven Drucker, MST-TR-98-52, October 1998.
<ftp://ftp.research.microsoft.com/pub/tr/tr-98-52.doc>

The final visualization from the Netscan project is a set of treemaps of relative volumes of postings/posters for large chunks of Usenet space. This element of the project is at a very early stage of development, and the examples shown here are some of the first attempts to map the whole of Usenet space.

Treemapping is an information visualization technique that can show a large hierarchical structure of data as a two-dimensional "space-filling" map. The technique was conceived in 1990 by Ben Shneiderman, a leading expert in human–computer interfaces and information visualization. In the treemap application, nodes (in this case, individual newsgroups) in the hierarchy are represented by rectangles sized proportionally to the number of messages each group receives and nested according to the lexical hierarchy of newsgroup names. A modified version of the treemaps approach was used in the Map of the Market. The example treemap opposite shows the whole of the Usenet space for the month of March 2000, and it represents 35,657 newsgroups with over 17 million messages. The examples below show significant portions of it in more detail: the *alt.tv.** and *comp.** hierarchies.

In the whole Usenet treemap the *alt* hierarchy is dominant, making up some 36 percent of all newsgroups and some 43 percent of all messages. Within the *alt* hierarchy itself, the *alt.binaries.** newsgroups are the most popular. The large undivided rectangles represent the huge super-newsgroups that receive tens of thousands of messages per month. The color-coding of the rectangles indicates the relative growth or decline of a newsgroup over the prior month, with dark-red coloring representing a steep fall in posting traffic, white being no change, and dark green indicating growth. In general, much of Usenet space is growing or stable, with only the odd smattering of red rectangles. Smith argues that the treemaps "resemble a land use map, with some areas seemingly more rural than others". Development of treemaps of Usenet is ongoing, and more revealing results are sure to emerge.

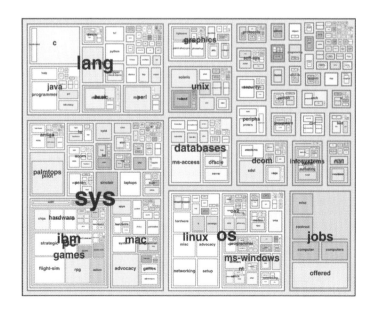

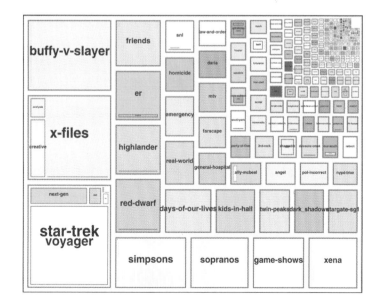

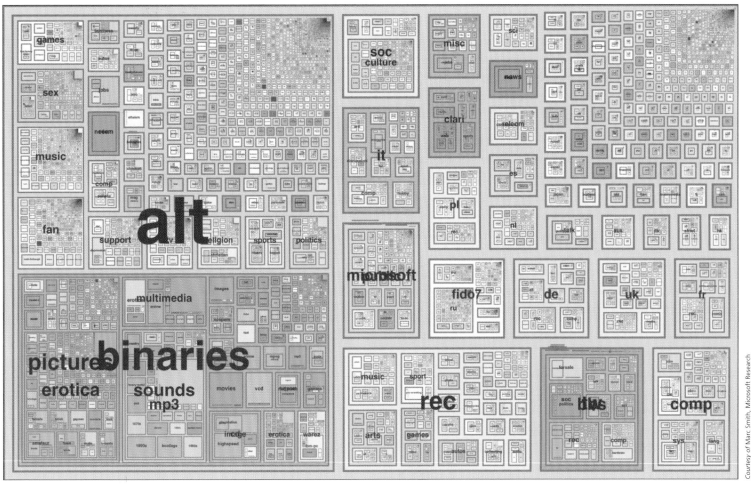

4.9: Netscan treemaps of Usenet space

chief cartographers: Marc A. Smith and Andrew T. Fiore (Collaboration and Multimedia Group, Microsoft Research).

aim: to show the relative volumes of postings/posters for large numbers of Usenet newsgroups. Part of the whole Netscan project for measuring and mapping the social geography of Usenet.

form: recursively nested rectangles representing a hierarchy of newsgroups, where the size of the area of the box shows the number of postings/posters.

technique: the treemap algorithm developed by Ben Shneiderman is a planar space-filling map, representing data hierarchically.

date: prototype images from November 2000.

further information: see <http://netscan.research.microsoft.com/>

further reading: for background information on treemaps, see "Treemaps for space-constrained visualization of hierarchies", by Ben Shneiderman, November 2000, Human–Computer Interaction Lab, University of Maryland, at <http://www.cs.umd.edu/hcil/treemaps/>

Mapping chat

The social media so far examined are asynchronous in nature. In this and the following sections, we turn our attention to synchronous media where interactions – although geographically dislocated – take place in real time in the same virtual space.

The first of these media we discuss is chat. Chat is an increasingly popular virtual medium through which individuals communicate by rapidly exchanging a series of short text messages, which form an ongoing conversation. Chat discussions are conducted within channels, meaning self-contained spaces that users can utilize to hold conversations. Many of these channels are themed, with users expected to conduct conversations on particular topics. One method adopted recently to try to explore the social relations and interactions between participants has been the spatialization of chat.

In conventional chat clients, the conversations are a multi-user experience where other people's dialog appears in the chat window, intermingling with that of each participant. Each sentence usually takes up one entry in the chat window, and the text can quickly scroll upward and off the screen as the conversation progresses. As such, chat is a very dynamic medium, comprising continuously flowing text that can be difficult to follow. Many first-time users can become easily confused, for it is difficult to follow multiple conversations that overlap and interweave, with sentences appearing out of sequence as people "talk" over the top of each other.

Chat Circles, developed by Judith Donath and her students, seeks to spatialize the scrolling reams of text, creating a graphical interface able to convey important structural information concerning the conversations as well as some of the unspoken nuances of face-to-face communication. The system comprises two distinct components, the first being the dynamic spatialization of Chat Circles and the second being an archival spatialization that enables a user to see and browse through an overview of conversations over time. Chat Circles employs simple, abstract two-dimensional graphics, encoding the key

chat characteristics of participant identity and conversational activity with the fundamental properties of shape, size and color for the representative elements.

The images display a range of example screenshots of the Chat Circles interface in action. The participants of chat rooms are represented as differently colored circles. All participants are represented, regardless of how much or how little they speak, enabling a user to see at a glance all those in the chat room. A major problem with conventional chat interfaces is that one is often only aware of the active speakers, as the text acts as a central indicator of presence. This makes people feel they must continually "speak" to maintain their presence, otherwise people will forget they are present. It also means one is not aware of "lurkers", who may be monitoring the conversation. Chat Circles aims to overcome this problem by its different format.

Another problem experienced by chat users is that chat space

4.10: Chat Circles

chief cartographers: Fernanda Viégas and Judith Donath, assisted by Joey Rozier, Rodrigo Leroux and Matt Lee (Sociable Media Group, Media Lab, MIT).
aim: to provide an enhanced chat client where it is easier to see the nature of the conversation(s) in the channel. Also, to provide a history of conversation.
form: bubbles of different colors represent chat-room participants, with the text appearing in the bubble. Bubbles can move to form distinct conversation clusters. The history of conversations is presented as vertically scrolling bars, somewhat like a seismograph trace.
technique: custom-written client software.
date: 1999. Screenshots taken in October 2000.
further information: see <http://chatcircles.media.mit.edu/>
further reading: "Chat Circles" by Fernanda Viégas and Judith Donath, in Proceedings of the CHI 99 Conference on Human Factors in Computing Systems, 15–20 May 1999, Pittsburgh, USA, pp. 9–16. <http://www.media.mit.edu/~fviegas/chat-circles_CHI.html>

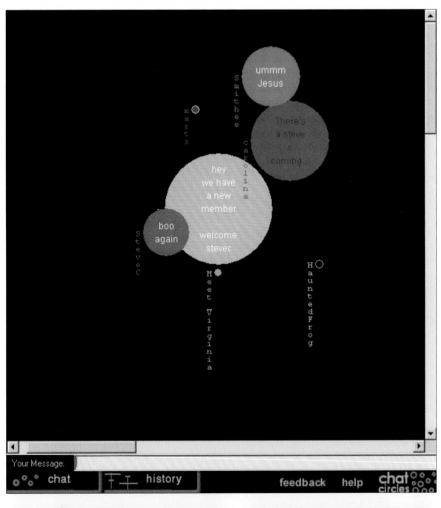

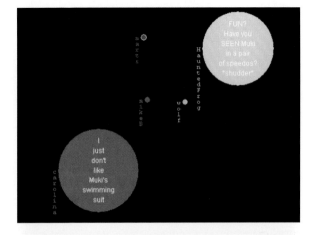

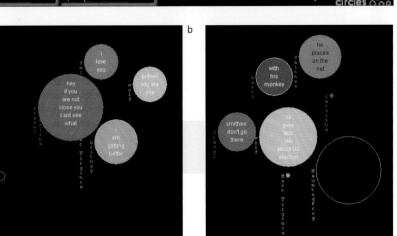

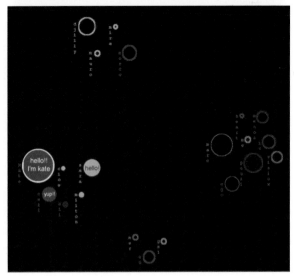

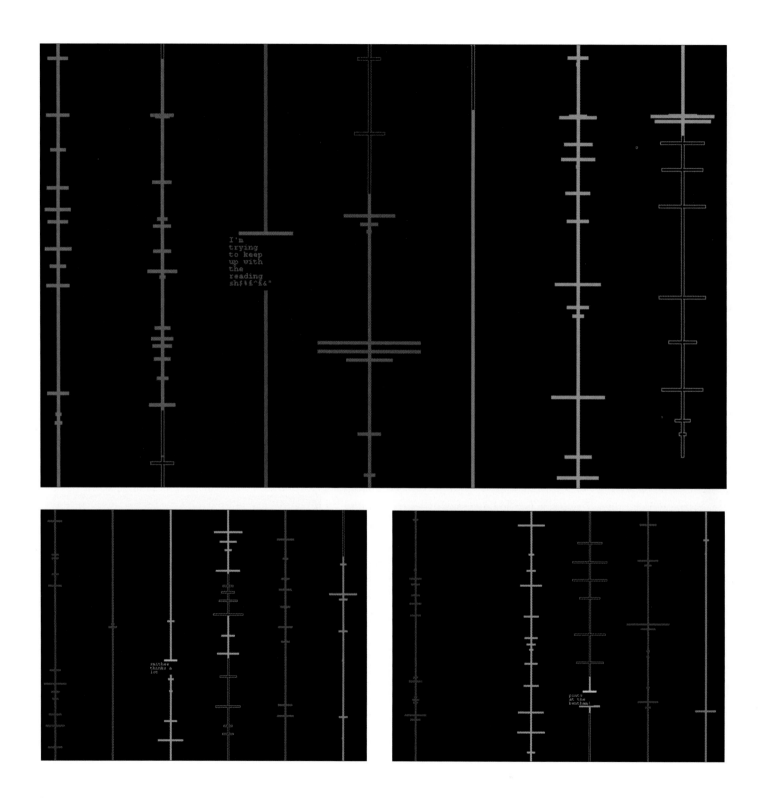

can become divided into a number of separate rooms or channels, each with small groups that are unaware of each other. Chat Circles overcomes this problem by displaying all the different rooms on a chat server on a single screen, dividing the chats into distinct conversational groups based on the spatial clustering of circles. However, it is only possible to "hear" the conversation in the limited geographic proximity of one's circle, so that no one is overwhelmed by a cacophony of chatter. Yet, crucially, it is possible to see the number and strength of the other conversations, indicated by the number and size of the circles. It also enables a user to move easily to another conversation by moving the appropriate circle to that cluster. Chat Circles thus employs spatial location in the interface as a tool for a kind of "geographic" (chat room) filtering. For example, you can clearly see conversational groupings in different areas of the screen, but a user should only be able to "hear" one set of conversations – that in the bottom left – and hence the text of the conversation is displayed in these circles.

The size and brightness of the circles is dependent on how much and how often people talk. A person's circle grows to accommodate the text of their speech. The circle then shrinks back and fades away, after each sentence, so the most active participants in the conversation are visually prominent on the spatialization, with large bright circles, while the lurkers are represented by small faded dots. The circle is also labeled with the person's name for easy identification and an individual's own circle is drawn with a white outline for enhanced distinctiveness. Over time, the dynamic of the conversation can be seen as the circles grow and shrink, and drift to different groups.

Unlike email messages or news articles, conversations in chat mode are rarely archived, and so once the conversation has disappeared from the window, it is lost. As such, the interactions in chat are highly ephemeral, and like the conversations at a party, unknown to those not intimately involved and often quickly forgotten. To accompany Chat Circles, therefore, a unique history function is also being developed, employing a timeline graphical metaphor somewhat reminiscent of a seismogram trace (opposite).

The spatialization produces what the creators call a "conversational landscape" tool, which enables the user to get an instant overview of the pulse conversation and easily browse individual chat contributions. Each participant is represented as a vertical line of the same color as the circles, with the z-dimension of the screen showing time. The horizontal bars are individual sentences, the width of the bars indicating the length of the sentence. The combination of all the lines shows the threads of the conversation. Browsing the threads is possible, with one simply pointing the mouse over a bar leading to the display of the actual dialog. For conversations that take place outside the "hearing" range of the particular participant, the sentences are shown by hollow horizontal bars. As with the main Chat Circles interface, it is a simple, minimalist approach to cyberspace spatialization, but it is potentially a powerful mapping of chat-space history – as its creators Viégas and Donath say, it "allows for a visualization of both group and individual patterns at the same time as it creates, by its mere shape and colors, a snapshot of an entire conversation in one image".

Paul Adams takes a different approach to spatializing chat, extending his work beyond chat rooms to include other modes of communication. His work is primarily interested in examining how transportation and communications technologies can "extend" the scope of the human body to reach out and interact across geographic space. In order to examine this, Adams has sought to model people's connections through time and space during the course of an ordinary day using three-dimensional time–space models created using the Vellum computer-aided design (CAD) application. The images opposite show one of these models, for one day in the lives of five interconnected people. Furthermore, the image below shows an illustration of the variety of means of communication.

The data for the model were gathered through detailed time diaries and interviews, recording the daily activities and social interactions (face-to-face, phone, letters, TV, radio, email) of a small group of connected people, who live in the Albany metropolitan area of New York. For each person, a separate 3-D model of his or her daily routine was built in the CAD package. The vertical axis represents time through the day and horizontal bars project out along the x-axis for different communications activities (such as making a phone call, watching television, talking face-to-face or sending an email). The horizontal length of the bar from the vertical axis shows the geographic distance of the activity, ranging from proximate face-to-face conversations to an international phone call. The x-y dimensions of the bars thus represent the scale of the activity in time–space, with the length of the bar being distance and the width of the bar being temporal. These individual time–space activity "bar charts" are combined into one model by arranging them evenly on the circumference of a circle. This enables a viewer to compare the shape and structure of the activities of the different people's daily lives as well as showing the communications links between them. So the curving between the different individuals represents interactions between them. For example, at the front of the model opposite, a link is drawn between Diann and Thomas that represents a face-to-face meeting between the two.

Adams argues that the 3-D models "reveal the existence of a kind of 'commuting' between physical and virtual places, an oscillation that occurs much faster than the older form of home–work commuting: every time one picks up a phone receiver, opens a book, or turns on the radio".

4.11: Models of human extensibility

chief cartographer: Paul Adams (Department of Geography, Texas A&M University).
aim: to model the communications between people.
form: 3-D time–space models.
technique: models are created in a computer-aided design (CAD) package.
date: 1999.
further information: see <http://geog.tamu.edu/faculty/adams/>
further reading: "Application of a CAD-Based Accessibility Model" by Paul Adams, in *Accessibility in the Information Age*, edited by Donald Janelle and David Hodge (Springer Verlag, 1999), pp. 217–239.

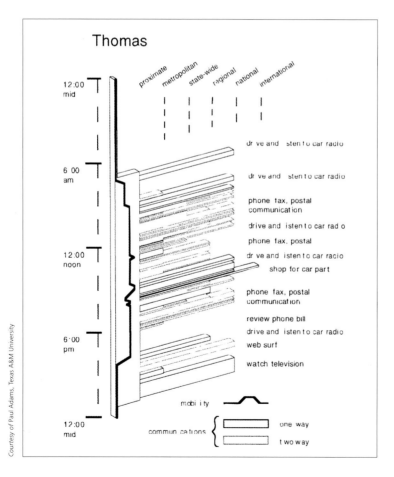

Courtesy of Paul Adams, Texas A&M University

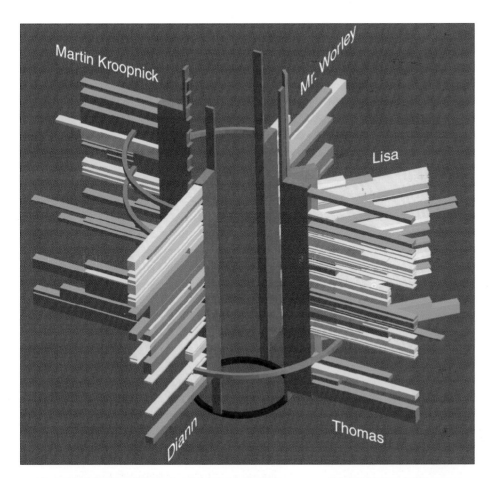

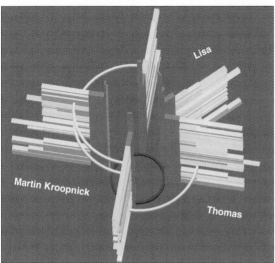

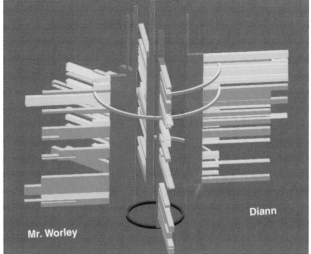

Courtesy of Paul Adams, Texas A&M University

Mapping MUDs

Multi-user dungeons (or domains), or MUDs for short, are a unique and often strange part of the Net. MUDs are virtual spaces created solely by written words; their space unfolds on the computer screen as scrolling text. MUDs are also social spaces shared by many players who are able to interact with each other and with the environment around them. The experience of MUDding (playing a MUD) is often described as like being inside a literary novel, but not merely as a reader. The MUD is a living novel, being written in real time by its players. (We are using MUD to refer also to MOOs (MUD object-oriented elements). MUD environments tend to be hard-coded, whereas players can change the environments of MOOs.)

A simple way in which to think about MUDs is to envisage them as chat rooms with an explicit geography. MUDs employ explicit spatial metaphors to create stages or scenes in which interactions are situated. So, instead of a conversation within a "spaceless" chat channel, interactions occur within a *place*, such as a living room, a bar, by the swimming pool or in a dungeon. Each room is known to its occupants by its *textual* description, and it can contain any number of objects (such as furniture). Rooms are linked by exits that enable MUD participants (generally known as players) to traverse from room to room. Some of the largest MUDs have many hundreds of rooms linked to form complex topologies. The topologies are often changing, with continual additions and subtractions of rooms. Much of this change is unplanned, the creative outcome of individual actions resulting in an evolving, organic structure. Such change can also lead to anomalies in the topological structures, the most obvious being "black hole" rooms that are only linked to the main MUD structure in one direction. So you can enter the room, but then have no means of exiting (except by direction teleportation to another location, if allowed, or by quitting and restarting). Some of these topological black holes are designed as deliberate pranks to trap unwary players, while other anomalies are simply mistakes in construction.

The linkages of the rooms into larger topologies of space, and the ability of players to move purposefully through them, travelling in distinct directions, all serve to create an approximate sense of spatiality. Given the explicit spatial topography of MUDs, one logical and analytical approach is to map its spatial form and extent. In this section we detail a number of attempts to do just that.

The history of MUDs, like much of cyberspace, can be traced to the experimental hacking of university students. The first MUD was created by two students, Roy Trubshaw and Richard Bartle, studying at Essex University in the United Kingdom at the end of the 1970s. Their first MUD, known as MUD1 or the Essex MUD, was written on the university mainframe in 1978 and it became available for networked users outside the university in the spring of 1980. Many of the concepts that lay behind Bartle and Trubshaw's first MUD had a much longer genealogy, owing a great deal to the dungeons-and-dragons computer games "Zork" and "Adventure" of the mid-1970s (although these were single-user games). These in turn were inspired by the "Dungeons and Dragons" board game created by Gary Gygax and Dave Arneson in the early 1970s – a complex role-playing game set in a Tolkienesque landscape of myth and fantasy, inhabited by warriors, elves, dragons and wizards.

The map over the page was drawn by Richard Bartle in 1983 and shows the layout of the approximately 420 rooms of MUD1. It is a simple black-and-white hand-drawn sketch map, but it is an important historical record of this part of cyberspace. The map comprises one large central section, with three smaller subsidiary maps that show details of clusters of rooms from the center of the MUD. Rooms are represented on the map by their name (or abbreviation) and the various connections between them are shown by the lines. Straight lines are simple "walkable" connections and are orientated according to the cardinal compass directions; stepped lines represent stairs; and curved lines are jumps. Each of the rooms shown on the map was envisioned for the players by a short textual description. To give a sense of what these were like, three descriptions – the *Badger's Sett*, the *Crow's Nest* and the *Shed* – are reproduced in the inset boxes shown.

The map was drawn several years after MUD1 was created and opened to players. The layout of rooms was transposed from the original "travel table", part of the MUD's core software code. Bartle made the map for the "wizzes" (Wizards). It shows the whole of the MUD's geography except for a few disconnected special rooms that were not accessible to normal players. MUD1 has had an extremely long life, in cyberspace terms, and it is still active after nearly 20 years. It ran at Essex University until September 1987 and then was licensed and operated by CompuServe. It was retired by CompuServe as part of its millennium-bug measures, and it now operates under the name British Legends (<http://www.british-legends.com/>). Bartle and Trubshaw are still actively designing and creating further MUDs with their company MUSE (standing for multi-user entertainment).

4.12: Sketch map of MUD1

chief cartographer: Richard Bartle (Muse Ltd, UK).
aim: to show the logical structure of the MUD by mapping the rooms and their interconnections. For use by MUD "wizards".
form: black-and-white sketch with names representing rooms and different line styles for the different types of connection. Linked to subsidiary maps that show complex areas of the MUD in detail.
technique: link-node topological map, hand-drawn with pencil and paper. Map derived from the "travel table".
date: 1983.
further information: see <http://www.mud.co.uk>; available nowadays to play as British Legends at <http://www.british-legends.com/>.
further reading: The Mud Connector <http://www.mudconnect.com/>
My Tiny Life: Crime and Passion in a Virtual World by Julian Dibbell (Fourth Estate, 1999).
<http://www.levity.com/julian/mytinylife/>.

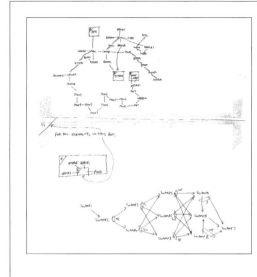

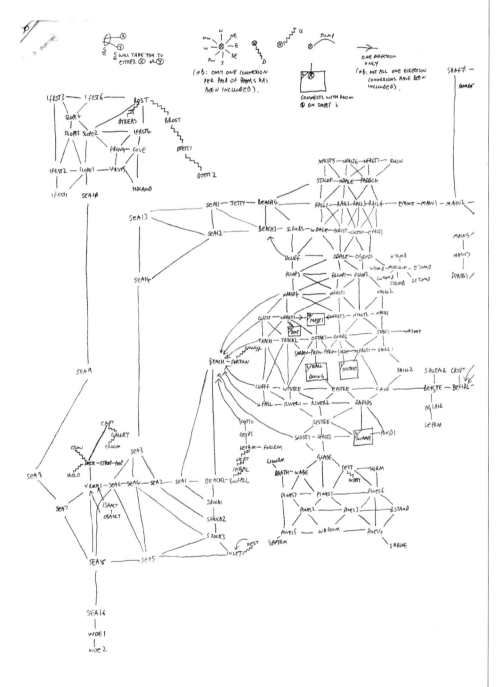

Crow's Nest

You are perched in the crow's nest, high above the sea where it juts out from the deck below. Far, far to the north-west can be glimpsed a mysterious island, and to the east over the sea are cliffs.

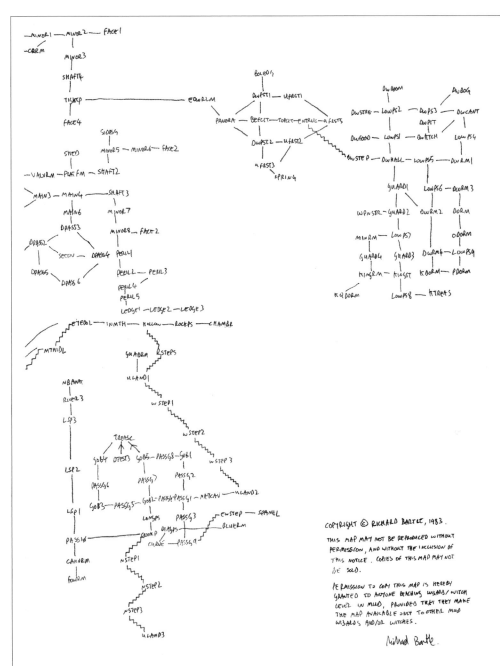

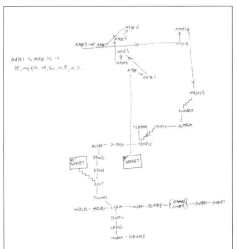

Badger's Sett

You are inside a small sett belonging to a badger. It looks as though the badger is hidden in a smaller hole above the entrance, but there is no conceivable way you can reach him. The room looks the ideal place to store objects, since the badger acts as guard. The only way out is up.

Shed

This is a decrepit but nevertheless rather sturdy shed, used by the tin miners in the old days to store their equipment. The way out is to the south, onto a platform at the top of a staircase.

A common way to create a simple map of a MUD is to use keyboard characters such as dashes and equals signs to draw cartographic features. The resultant maps are often called "ASCII text maps". Perhaps the fact that they are just text, like the MUD itself, appeals to many of their cartographers. They can also be created in a basic text editor or word processor without any graphics software. Moreover, they can be viewed on any computer (as long as they are displayed with a fixed-width font), again without any special software. Countering this is the fact that ASCII maps can be extremely time-consuming to create, requiring a particular degree of patience for large ones; furthermore, there is a limit to the type and scale of graphical forms that can be created just with keyboard characters.

The example opposite is from PhoenixMUD (a combat-orientated MUD started in March 1995, with over 8,000 rooms). The map was created to provide an overview of the geography of the continent of Caledon, one of three continents that make up the world of PhoenixMUD. (Maps of the other two continents and cities are available.) The map is at the regional scale, showing general areas such as cities, forests, deserts, and their relative geographic positions. It does not show detailed room topologies (unlike Bartle's MUD1 map). Caledon contains the major city of PhoenixMUD, namely Heliopolis, which is where players begin their adventures. According to the description accompanying the map, "the area is ancient, as evidenced by the many ruins of long-forgotten civilizations, including Old Thalos and Rhyodin. Numerous castles, towers, and natural attractions await the eager explorer".

4.13: ASCII-style map of part of PhoenixMUD

chief cartographer: not known.
aim: to show the geography of the continent of Caledon, the main area of PhoenixMUD.
form: created with ASCII characters such as dashes and equals signs.
technique: manual ASCII plotting.
date: 1998.
further information: see <http://www.phoenixmud.org/>

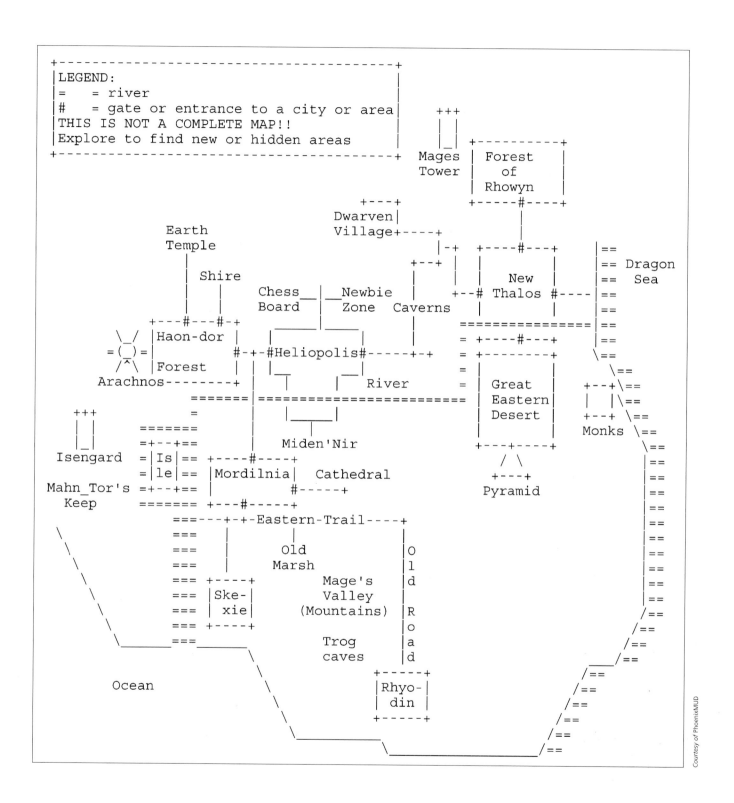

Courtesy of PhoenixMUD

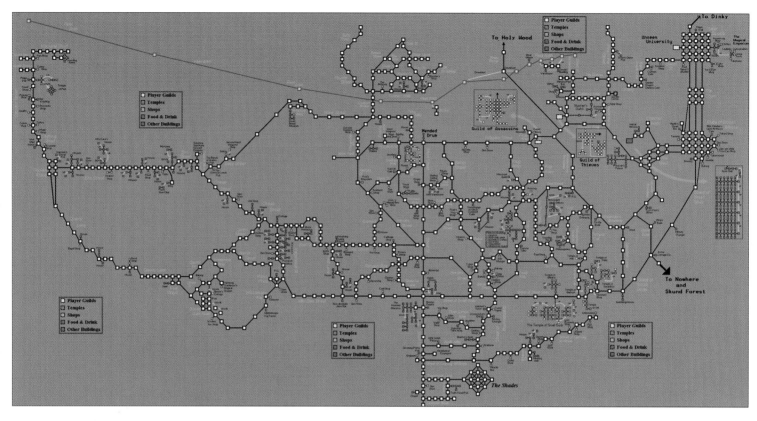

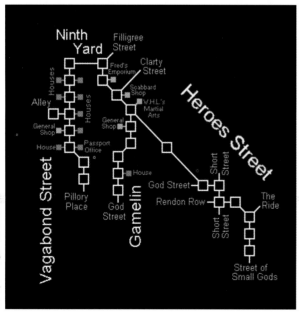

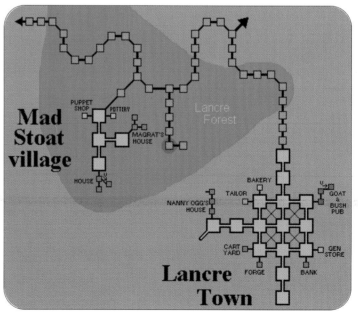

Courtesy of "Morgoth"/Daniel Staniforth

Discworld MUD is based on the hugely popular fantasy world of Terry Pratchett's novels. The MUD is long-running, large and popular, with between 150 and 200 players logged in at any one time. The geography of the Discworld MUD has also been well mapped by various players. The most comprehensive set of maps was produced by player "Choppy", and they are currently maintained by "Morgoth" (Daniel Staniforth). These form the atlas of Discworld, showing the complex room topologies from regions and cities down to individual streets and building layouts. The maps are high-quality in terms of cartographic design, and they represent many hours of dedicated volunteer work. They are perhaps the most complete mapping of any MUD.

Opposite are three examples from the Discworld Atlas. The maps use a familiar link–node graphical style to portray detailed room topologies. Individual rooms are represented by small boxes, with significant ones colored by function and often labeled. Black lines between the squares indicate possible pathways. The large map shows the geography of Ankh Morpork, "contayning withyn all thye streets of thye Great Citye of thye Dysc". The city is large, with many rooms organized into a complex pattern of interconnecting streets. There are a large number of shops, houses, temples and so on. The blue line across the top is the River Ankh. A number of small inset maps show the detailed room layouts inside important player guild buildings. More detailed maps of all the streets and key buildings in the city of Ankh Morpork are also available. An example of Heroes Street is shown bottom-left. This street is in the center of the main Ankh Morpork map. The small brown boxes are houses and the green ones are shops. The diagram shown bottom-right plots the outlying town of Lancre and Mad Stoat Village, which are in the Circle Sea region, close to the Ramtop Mountains.

4.14: Maps from the Discworld Atlas

chief cartographers: original map research and drawings by player "Choppy", now maintained by player "Morgoth" (Daniel Staniforth).
aim: to show the geography of the Discworld MUD by mapping room topologies at different scales, from regional to local streets and building layouts.
form: well-designed map with squares representing rooms. Clear color-coding and labeling to identify significant locations (temples, guilds, shops, etc.).
technique: hand-drawn sketches made on graph paper whilst exploring the MUD. These have then been used to create finished maps using standard graphics software.
date: 1998.
further information: browse the Discworld Atlas at <http://discworld.imaginary.com>

FurryMUCK is based around the role-playing theme of anthropomorphics, where players role-play as human-like but animal (furry) characters. Like the Discworld MUD, FurryMUCK regularly has over 150 players online, and the maps of the domain serve as a useful tool for newcomers and regulars alike. The example shown below is one of a series of ten maps maintained by player "Quill" (Graham J. Clarke), based on original research and mapping by Tom Turrittin in 1994. The maps only show a small portion of FurryMUCK – the central areas that can be reached on foot. The example chosen, Map 5, shows the Far North of the world, including Eagle Mountains and the Arctic. The graphical style of the map is somewhat like a circuit diagram, with circular nodes and straight lines showing the room topologies. The rooms are laid out on a grid to give a regular structure.

Abandon All Hope MUD is a multiclass role-playing MUD where players progress through a series of roles and levels. As their rating improves, the classes (e.g., thief, magus, cleric, warrior, etc.), skills and areas available to them increase.

This MUD was mapped by Arkady in March 1999 using a stylized dot map somewhat reminiscent of a star chart. Rooms

are represented by small dots, many of which are labeled to indicate interesting locations. Clusters of rooms are coded using the same color scheme. The rooms are arranged spatially to show the relative locations of rooms but, unlike the previous MUD maps we have looked at, Arkady's map does not show how the rooms are linked. The map is partially interactive, in that you can zoom into particular areas of interest, but no further detail is revealed.

Many more maps of Abandon All Hope MUD are available as ASCII-style sketch maps.

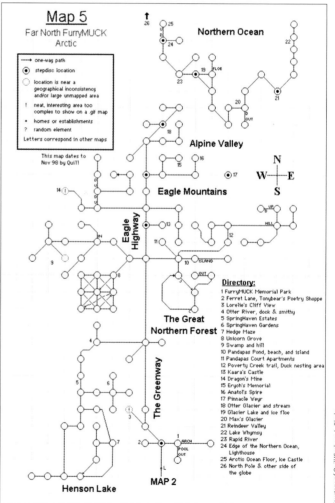

Courtesy of Quill/Graham J. Clarke

4.15: FurryMUCK Map

chief cartographers: player "Quill" (Graham J. Clarke), based on original maps from Tom Turrittin.
aim: to map the room topology of the central areas of FurryMUCK that are accessible by walking.
form: black-and-white wiring-style diagram. Rooms represented by circles, with interesting and significant locations listed in a directory.
technique: hand-drawn maps.
date: November 1998.
further information: maps available at <http://whales.magna.com.au/furry/maps/> access FurryMUCK at <http://www.furry.org>

4.16: Map of Abandon All Hope MUD

chief cartographer: Arkady.
aim: to show the relative geography of the rooms of Abandon All Hope MUD.
form: clustered colored dots to represent rooms.
technique: digital map in Shockwave.
date: March 1999.
further information: maps are available at <http://www.grayarea.com/mudmaps.html> access Abandon All Hope MUD at <http://www.grayarea.com/mud.htm>

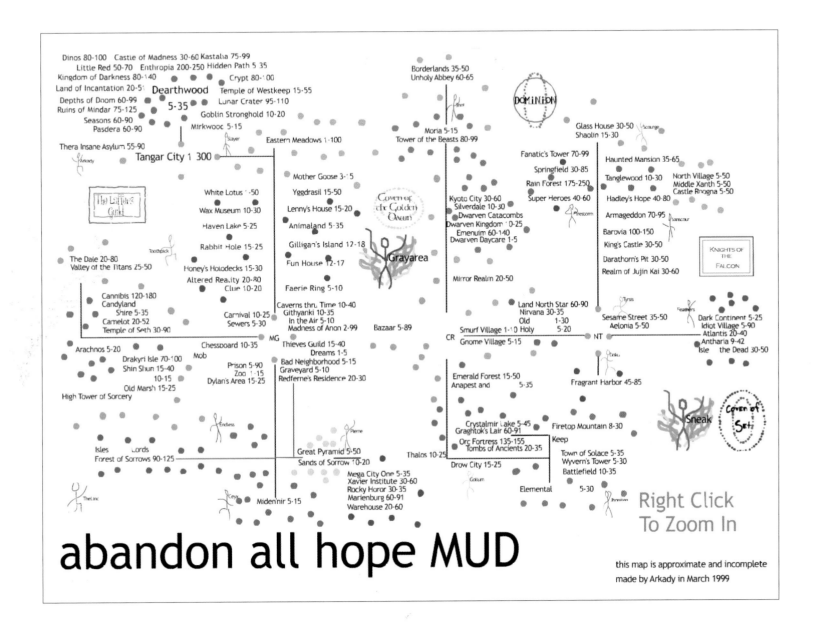

Dinos 80-100 Castle of Madness 30-60 Kastalia 75-99
Little Red 50-70 Enthropia 200-250 Hidden Path 5-35
Kingdom of Darkness 80-140   Crypt 80-100
Land of Incantation 20-51
Depths of Doom 60-99   **Dearthwood**   Temple of Westkeep 15-55
Ruins of Mindar 75-125   **5-35**   Lunar Crater 95-110
Seasons 60-90   Goblin Stronghold 10-20
Pasdera 60-90   Mirkwood 5-15

Borderlands 35-50
Unholy Abbey 60-65

DOMINION

Thera Insane Asylum 55-90   Eastern Meadows 1-100

Moria 5-15
Tower of the Beasts 80-99

Glass House 30-50
Shaolin 15-30

Tangar City 1-300

The Lifters Guild

Mother Goose 3-15

Fanatic's Tower 70-99
Springfield 30-85
Rain Forest 175-250
Super Heroes 40-60

Haunted Mansion 35-65

Tanglewood 10-30   North Village 5-50
Middle Xanth 5-50
Castle Rhogna 5-50

White Lotus 1-50
Wax Museum 10-30
Haven Lake 5-25
Rabbit Hole 15-25

Yggdrasil 15-50
Lenny's House 15-20
Animaland 5-35
Gilligan's Island 12-18

Coven of the Golden Dawn

Kyoto City 30-60
Silverdale 10-30
Dwarven Catacombs
Dwarven Kingdom 10-25
Emenuim 60-140
Dwarven Daycare 1-5

Hadley's Hope 40-80

Armageddon 70-95

Barovia 100-150
King's Castle 30-50

KNIGHTS OF THE FALCON

The Dale 20-80
Valley of the Titans 25-50

Honey's Holodecks 15-30
Altered Reality 20-80
Clue 10-20

Fun House 12-17

Faerie Ring 5-10

Grayarea

Mirror Realm 20-50

Darathorn's Pit 30-50
Realm of Jujin Kai 30-60

Cannibis 120-180
Candyland
Shire 5-35
Camelot 20-52
Temple of Seth 30-90

Carnival 10-25
Sewers 5-30

Caverns thru Time 10-40
Githyanki 10-35
In the Air 5-10
Madness of Anon 2-99

Bazaar 5-89

Land North Star 60-90
Nirvana 30-35
Old   1-30
Smurf Village 1-10 Holy 5-20
Gnome Village 5-15

Sesame Street 35-50
Aelonia 5-50

Dark Continent 5-25
Idiot Village 5-90
Atlantis 20-40
Antharia 9-42
Isle the Dead 30-50

Arachnos 5-20

Chessboard 10-35
MG
Mob

Thieves Guild 15-40
Dreams 1-5
Bad Neighborhood 5-15
Graveyard 5-10
Redferne's Residence 20-30

CR

Drakyri Isle 70-100
Shin Shun 15-40
10-15
Old Marsh 15-25

Prison 5-90
Zoo 1-15
Dylan's Area 15-25

Emerald Forest 15-50
Anapest and   5-35

Fragrant Harbor 45-85

High Tower of Sorcery

Sneak

Coven of Seth

Isles   Lords
Forest of Sorrows 90-125

Great Pyramid 5-50
Sands of Sorrow 10-20

Crystalmir Lake 5-45
Graghtok's Lair 60-91
Orc Fortress 135-155
Tombs of Ancients 20-35

Thalos 10-25

Firetop Mountain 8-30
Keep

Town of Solace 5-35
Wyvern's Tower 5-30
Battlefield 10-35

TheLink

Mega City One 5-35
Xavier Institute 30-60
Rocky Horor 30-35
Marienburg 60-91
Warehouse 20-60

Miden'nir 5-15

Drow City 15-25

Elemental   5-30

**Right Click
To Zoom In**

abandon all hope MUD

this map is approximate and incomplete
made by Arkady in March 1999

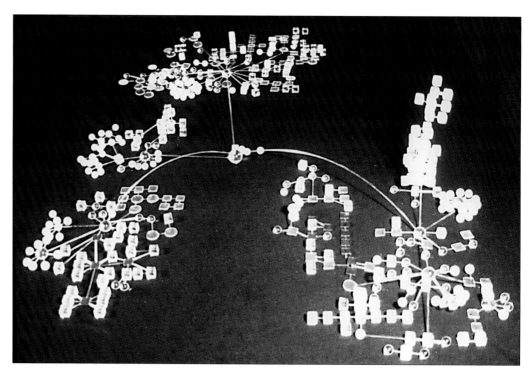

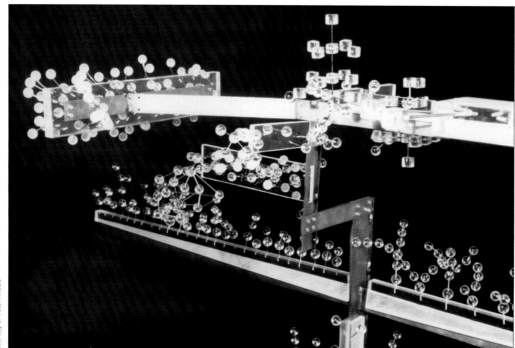

These two example MUD maps are somewhat different from the others we have examined thus far. They were in fact physical models of MUD room topologies rather than two-dimensional sketches or diagrams, created by Peter Anders and his students in the fall of 1995. Working in pairs, students conducted a detailed field survey of specific MUDs, noting and sketching the geography in a notebook as they explored. Next, they constructed what Anders terms "logical adjacency models" – physical three-dimensional models built with Plexiglass cubes (represents rooms) and rods (connections) that visually graph the MUD structure, looking like a molecular model commonly used in chemistry or biology. As far as possible, the arrangement of cubes and rods is positioned spatially congruent to the MUD structure. Rooms that are connected – but not connected by Euclidean geometry (e.g., they are accessed by teleporting) – are represented by spheres positioned arbitrarily. Only the publicly accessible geography of the MUD has been mapped, and so private spaces have been omitted. In all, ten MUDs have been modeled, and the images opposite show two examples: the logical adjacency model of BayMOO produced by Thomas Vollaro and Susan Sealer (top) and the model of MediaMOO created by George Paschalis and Michael Lisowski (bottom).

The BayMOO model has three distinct zones joined by a central node known as the Aquatic Dome. These clusters of rooms in the map are the three major areas of BayMOO, all with distinctive characteristics, known as "The Bay Area", "Netspace" and "Other Worlds". Anders says the stick-and-ball LAMs (logical adjacency models) developed by his students reveal the distinct structure of a MUD from the topology of its rooms, much like a fingerprint provides unique identification of a person. The fingerprint of a particular MUD is determined to a large degree by the political structure of the MUD, and Anders says:

MUDs whose maps resemble an orthogonal grid of cubic rooms reflect a strong administration of wizards – a top-down control of construction in the domain. On the other hand, in democratic, bottom-up managed MUDs, users are free to build spaces without constraint. LAMs of these MUDs tend to be shaggy clusters of spheres, as the directional grid is not followed rigorously.

Anders continues to explore the spatial structures of cyberspace. He has published a book entitled *Envisioning Cyberspace* (McGraw-Hill, 1998), in which he delves deeper into this fascinating topic. In his current doctoral research he is developing "cybrid space, which combines attributes of physical and electronic spaces".

4.17: Physical 3-D models of (a) BayMOO and (b) MediaMOO

chief cartographers: under supervision of Peter Anders, the BayMOO model was created by Thomas Vollaro and Susan Sealer; the MediaMOO model was created by George Paschalis and Michael Lisowski (School of Architecture, New Jersey Institute of Technology).
aim: to show room topology of large MUDs.
form: models in three dimensions using spheres and cubes to represent rooms. Has the appearance of a molecular model.
technique: physical, Plexiglass model.
date: Fall 1995.
further information: Peter Anders' homepage is at <http://www.mindspace.net/> try also BayMOO at <http://baymoo.org:4242/> and more information on MediaMOO is available at <http://www.cc.gatech.edu/~asb/MediaMOO/>
further reading: *Envisioning Cyberspace: Designing 3D Electronic Space*, by Peter Anders (McGraw-Hill, 1998).

All the maps of MUDs that we have looked at so far chart the "physical" topologies of the virtual environment. This example, however, takes a different approach, mapping the complex topologies of the *social* connections between players. The huge and dense mesh of connections shows the social geography of LambdaMOO by mapping how over half of the 4,800 or so players relate to each other. LambdaMOO is a well-established and well-known virtual environment. It was created at Xerox PARC by Pavel Curtis in 1990.

The map was created using social statistics gathered by Cobot, a software agent that has "lived" in LambdaMOO since June 1999. It sits in the "living room" and observes the social interactions of players. Cobot was created by Michael Kearns and Charles Lee Isbell Jr and colleagues as an AI research project at AT&T Labs. They state that Cobot's "goal is to interact with other members of the community and to become a vital, useful and accepted part of its social fabric". Cobot is a kind of software sociologist, building detailed statistical profiles of everyone's social interaction in terms of verbal communication and also the rich MOO vocabulary of non-verbal emoters such as hugs, smiles and shrugs. In one year, Cobot has recorded over 2.5 million events. Importantly, players can also talk to Cobot, and in particular they can ask questions about their sociability. For example, Cobot uses its database of social statistics to give an empirical answer to questions such as "Who are my playmates?", "Who acts like me?" and "Who loves me?" As a result, Cobot has become a very popular fixture in LambdaMOO.

The maps were produced in an effort to visualize and begin to understand the complexity of social relations in a large, vibrant social cyberspace and to see whether cliques and small-world social networks came to exist. The graphs are created by Christian Shelton using "dot", an application capable of handling large graphing tasks (developed by researchers Eleftherios Koutsofios and Stephen North, also of AT&T Labs). The maps were originally designed for plotting as large wall-sized 6 ft by 4 ft posters, and so those on the next two pages have had to be shrunk greatly to fit the pages of this book.

In the graphs, each player is represented as a small elliptical node. The graph is anonymized, so that the node contains a code number rather than an actual player's name. The nodes are color-coded according to the kinds of communications player exhibit, measured in two dimensions: amount of interaction and the amount of emoting. Red–purple colors tend to be players with an average degree of interaction through "speaking"; yellow, normal through emoting; blue, heavy interactors through speaking; green, heavy interactors through emoting. There are some 2,784 players shown in the full map, as well as many thousands of connections gathered over 14½ months by Cobot. On average, a player is connected to 1.3 other players. The graph opposite shows the out and in connections for each player, with an arrow from each player to all his or her playmates. The most connected player is 4841, a hacker who has inflated his position by writing code that can generate messages to all other users online. This is followed by Cobot itself, which has 288 connections. Another highly connected player is HFh – Charles Lee Isbell, one of Cobot's creators – with 151, but soon below this the number of out-connections drops off rapidly.

4.18: Cobot map of the social structure of LambdaMOO

chief cartographers: Charles Lee Isbell Jr, Michael Kearns, Christian Shelton and colleagues (AT&T Research, Shannon Labs).
aim: to show the complex social connections between all LambdaMOO players, based on their verbal and non-verbal interactions.
form: a massive and complex graph, where players are represented by small colored ellipses. Originally designed as a large wall-sized plot.
technique: graph based on social statistics gathered by the Cobot software agent, which "lives" in LambdaMOO. Graphs drawn using the "dot" application.
date: November 2000.
further information: see <http://cobot.research.att.com>
further reading: *Cobot in LambdaMOO: A Social Statistics Agent*, by Charles Lee Isbell Jr., Michael Kearns, *et al*. (AAAI, 2000).
<http://www.research.att.com/~mkearns/papers/cobot.pdf>.

Courtesy of Charles Lee Isbell Jr and Michael Kearns, AT&T Research

Clearly, it is difficult to read and study the whole graph of LambdaMOO's social geography (especially given the reduction to fit the page). As a consequence, the first graph should be viewed as an overview, displaying global patterns of complexity. To explore the social geography further, a number of interesting subgraphs have been created, and two are shown here. The first (bottom-left) shows the 368 players who are directly connected with Cobot or HFh. Interestingly, not everyone is clustered close to these two players and there are other distinct social cliques evident. The second example shows the 11 most connected (and thus "popular") people.

The development of Cobot continues, including making it/him proactive so that it/he can initiate interactions with other players rather than just respond to direct questions. Cobot's creators say that "the hope is to build an agent that will eventually take unprompted actions that are meaningful, useful or amusing to users".

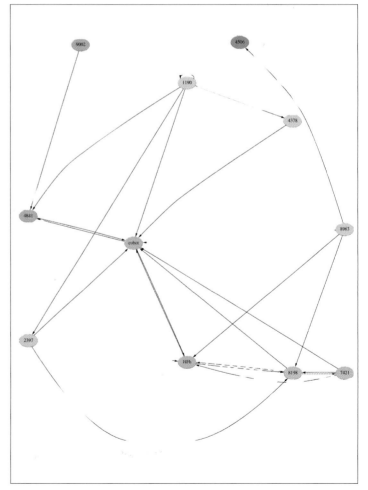

Mapping virtual worlds

Virtual worlds are very much like MUDs; but rather than interaction being mediated solely by text, a graphical interface exists that reveals visually the physical landscape and the representations of the participants. A number of such worlds now exist and include AlphaWorld, V-Chat, InterSpace, Worlds Chat, WorldsAway, The Palace, Deuxième Monde, CyberGate and Online Traveller. Each system aims to provide a visual, spatial domain (using 2-D, 2.5-D or fully 3-D graphics) that can be shared by many people socializing in real time. At present, they are perhaps the closest form of online interaction to the shared, immersive, VR worlds envisaged by cyberpunk writers (see chapter 5). Because they are quite clearly spatial in nature, a number of attempts have been made to map their geographic extent. Here, we detail a number of different attempts to map the spatial extent of AlphaWorld – probably the most popular virtual world and certainly the most studied and mapped since it went online in 1995.

The original geography of AlphaWorld was a flat, featureless plain, with no natural features, that stretched for hundreds of virtual kilometres in every direction. This plain was colored a uniform shade of green to signify that it was virgin territory waiting to be claimed. The total area of the flat plain is exactly 429,025 square kilometres, which is 43 percent larger than the United Kingdom or 4.4 percent larger than California. Unlike the UK, the borders of AlphaWorld are straight, forming an exact square of land 655 kilometres across. A Cartesian coordinate system is used to delineate this space, with an origin point (0,0) located at the dead center of the world. This center point is known as Ground Zero (GZ) and acts as a focal point – the point at which most people enter AlphaWorld. As a result, the area around Ground Zero has the greatest density of development, containing the oldest structures in AlphaWorld; it is also the most densely populated location. The coordinate system is important because it divides the plain into a series of 10×10 meter cells and allows people to navigate, by means of teleportation, to any point in AlphaWorld using an (x,y)

coordinate address. For example, coordinates (67N, 42W) translate to 670 meters north and 420 metres west of GZ.

Since AlphaWorld came into existence, inhabitants have been busy claiming land and building all manner of structures, from modest suburban-style homes to grand castles. As of November 2000, 67.7 million objects had been placed by the inhabitants. All the objects used in construction, such as windows, doors, stairs and furniture, are appropriately scaled in relation to representational size, which is limited and constant. The maps of AlphaWorld are interesting because they can be used as a kind of proxy for analyzing the underlying processes of urban and social development. In some senses, we can infer a lot of the nature of the virtual community by studying what is built and where building takes place.

We start our discussion by examining how land is claimed and structures are built. Further information on AlphaWorld can be found at <http://www.activeworlds.com>. For a general overview of virtual worlds on the Internet, see Bruce Damer's book *Avatars! Exploring and Building Virtual Worlds on the Internet* (Peachpit Press, 1997).

AlphaWorld enables users to "own" land and design and build homesteads, thereby constructing their own places for social interaction. This "homesteading" facility was a conscious aspect of software design and is unique amongst competing commercial Internet virtual worlds. It has been enthusiastically grasped by many thousands of people since AlphaWorld was opened.

The first step in building is to locate a plot of empty land that is not owned by anyone else. This can be difficult near GZ because of the density of existing urban development; however, there is still plenty of virgin land further from the center. Claiming land consists simply of building in situ, and there is no limit as to how much land can be claimed. It is important to claim all necessary land, however, as others can build on virgin territory and thus in the middle of a development. This in fact gives rise to one of the major sources of conflict in AlphaWorld: building disputes.

Building is undertaken with predefined objects, much like virtual Lego bricks, such as road sections, wall panels, doors, windows, flowers and furniture. In total there are over 1,000 different objects available and they can be put together piece by piece to create larger structures. The Active Worlds browser provides rudimentary tools to select and manipulate the objects so as to put them in a desired position. Construction of large buildings, using hundreds of individual objects, requires a considerable amount of skill, effort and – above all – patience.

As with a Lego set, buildings can only be built with the pieces provided, and it is not possible to create your own objects in AlphaWorld. This means that the built environment of AlphaWorld can have a somewhat homogenous appearance. Despite this limitation, the individual creativity of the citizens has flowered, with all manner of interesting structures having been constructed. Some are well designed and aesthetically pleasing, but there are also equal measures of ugly and half-finished structures. Here, we present a range of architecture to give a flavour of some of the structures built. The buildings are located in relation to maps of AlphaWorld.

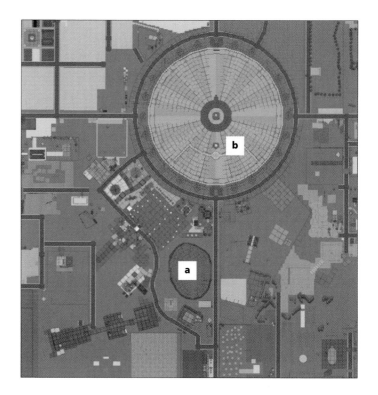

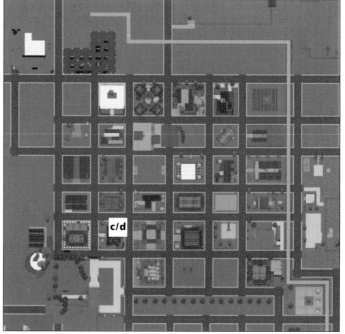

a Colosseo [location 224 S, 104 W]

The "colosseo" is a large Romanesque monument built by a tourist in October 1996.

b Greatlight Keep [location 202 S, 105 W]

Greatlight Keep is a large stone tower at the heart of a huge circular sundial-type structure. It was built by user "LittleBull", a prominent early builder in Active Worlds, in August 1996.

c and d Mock Tudor House [location 1635 N, 2397 W]

Two views of a well-designed British mock-Tudor suburban detached house at "77 Rockinghorse Way". It was built by user "Tranthum" in December 1996. The interior, however, is empty. This house is part of the planned development, started in 1995, called "Suburbia", with a grid of roads and large detached houses.

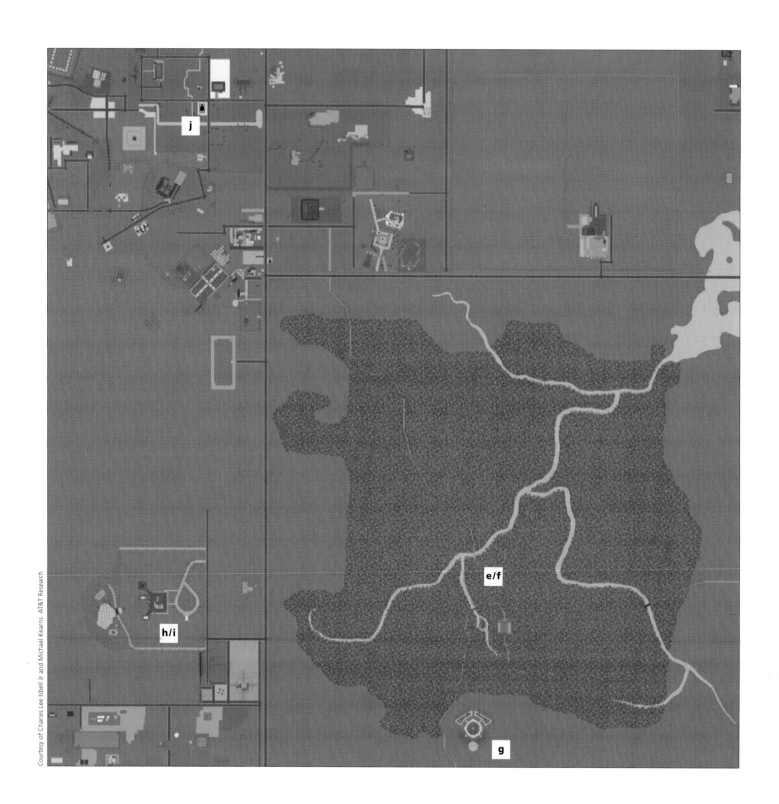

e–f The Great Forest and surroundings [location 598 s, 565 E]

g [location 635 s, 566 E]

The Great Forest was built between November 1995 and spring 1996, also by "LittleBull". It is one of the largest, most impressive building projects in AlphaWorld, containing tens of thousands of trees. A river and tributaries flow through the heart of the forest, crossed by a footbridge [location 598 S, 565 E]. There is also a small ornamental garden on the edge of the forest, with a large chess set, also built by "LittleBull" in March 1996 [location 635 S, 566 E].

h and i Pimlico Palace [location 597 S, 465 E]

To the west of the Great Forest is Pimlico Palace, a large, modern-style country house built by user "KomodoDragon" in March 1997. It has a minimalist interior décor and nice gardens, with a pool. It is set in substantial grounds, including a lake, and is approached from the main road via a sweeping driveway.

j Douggie's House [location 446 S, 482 E]

A compact and futurist-looking house built by user "Douggie" in November 1996.

Monorail System

The monorail system built by user "Ko Toff" in February 1998 is an immense project, spanning an area of the order of 27 square kilometres. It includes more than 40 stations and 8 distinct loops of track. The image shows one of the stations.

Titanic [location 2349 N, 1917 W]

A vast and very well-designed model of *Titanic* in a dry dock in the middle of AlphaWorld. It was built by user "Omega" in July 1999.

Pieter's Dutch Castle [location 52 S, 109 W]

A grand stone castle built by Pieter van der Meulen in April 1998. The castle is surrounded by a deep moat and a working drawbridge.

These striking black-and-white maps show the density of urban development for the whole 429,025 square kilometres of AlphaWorld's virtual land at specific times over recent years. The maps have been created by Greg Roelofs and Pieter van der Meulen as part of their comprehensive AlphaWorld mapping project. In the maps, the density of building is indicated by the brightness of the pixels. Bright white blobs are therefore congruous with towns, areas speckled with white dots are largely "rural" with scattered homesteads, while solid black areas remain undeveloped "wilderness". The maps look like satellite maps of the Earth taken at night when major cities are visible because of the light they are emitting.

The main image (opposite, top) is the latest map at the time of writing, showing the world as at 12 June 2000. It is clear from this image that the most developed area of AlphaWorld is the densely built city around Ground Zero (GZ). Given its importance as the entry point into AlphaWorld, the area around GZ has become heavily urbanized, with roads and buildings of all shapes and sizes sprawling out in many directions for many kilometres. (More detailed maps of GZ city are given on pages 205–7).

The most obvious feature of the urban geography revealed by this map are the straight ribbons of development that project out from GZ along the eight principal compass axes. Together, these ribbons form a distinctive star shape. Towns and other small settlements lie along these axes, looking like bright beads strung along a necklace. The directional structure of urban development is largely the result of the single entry point and coordinate system, with people choosing regular and memorable coordinates, such as [50 N, 50 W] or [1555 E, 1555 S] as the location for their homestead. Once a pioneer has started building, other citizens will build alongside, either by invitation or just to be close to another settlement.

The smaller images show density maps from three earlier phases of development, taken in 1997, 1998 and 1999 respectively. Looking at this sequence of maps, one can clearly see the rate of urban development in AlphaWorld. Over time the city at GZ has sprawled further from GZ, filling in spaces and extending along the principal coordinate axes. The star pattern is clearly evident, even in the earliest map from July 1997, and this has become stronger over time. One of the reasons for the sprawl effect is the fact that land cannot be redeveloped even if redundant – only the builder of a property can alter that property. New developers are thus forced to build on virgin land.

4.20: Density of urban development in AlphaWorld

chief cartographers: Greg Roelofs and Pieter van der Meulen (Philips Silicon Valley Center, Sunnyvale, California). Part of the Vevo mapping project that produces very detailed maps of the urban geography of AlphaWorld on a quarterly basis.
aim: to show the urban density for the whole of AlphaWorld at a snapshot in time.
form: 2-D black-and-white map, looking like a satellite image of the Earth taken at night, where the cities stand out because of their lights.
technique: the brighter the pixel in the image, the more buildings at that location. Generated from the AlphaWorld building database using custom software.
dates: main image shows AlphaWorld, 12 June 2000. Smaller side images are from 31 July 1997, 3 June 1998, and 16 August 1999.
further information: see <http://awmap.vevo.com>

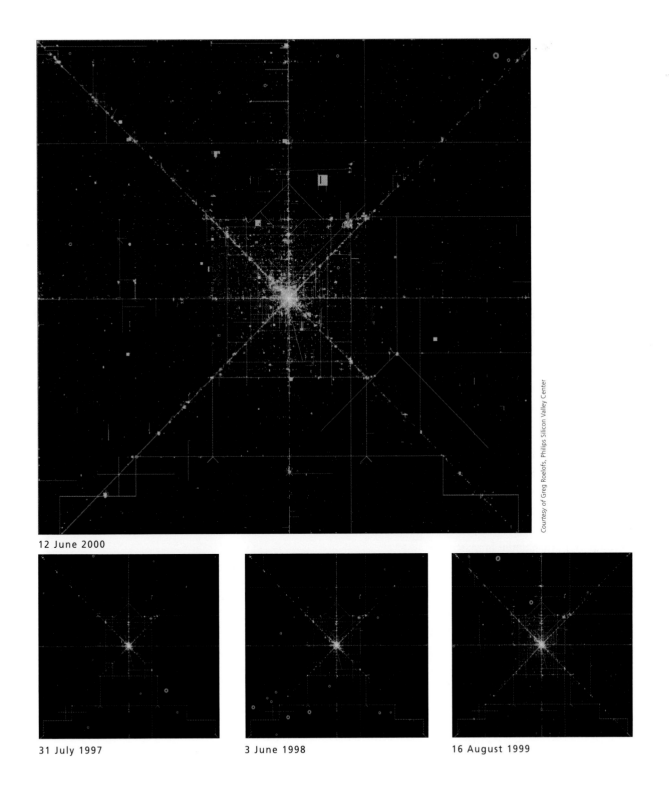

12 June 2000

31 July 1997

3 June 1998

16 August 1999

What would the urban morphology of a virtual city be like? This sequence of "satellite" maps of AlphaWorld's GZ City provides us with a relatively good idea. The maps show the growth of buildings, streets, parks and gardens, as designed and constructed by the citizens of AlphaWorld over the past five years. The cartographer responsible is Roland Vilett, the lead programer on the virtual world system at Activeworlds.com, which includes AlphaWorld. These maps are at a much more detailed scale than the density maps examined in Map 4.20, only covering the GZ City area which amounts to some 400 square kilometres. (This is just 0.3 percent of the total expanse of AlphaWorld.)

The maps on the next three pages reveal several interesting features of the AlphaWorld's urban geography. The first, and most obvious, impression is just how much the map looks like a map of a city in the real world. Each has the seemingly organic complexity and disorderliness of real cities, which grow over time through many thousands of individually planned actions. This likeness is clearly apparent if Vilett's maps of AlphaWorld's city are compared with remotely sensed images at a similar scale to real-world cities.

The extent of urban growth, spreading out from GZ, is clearly apparent when the maps are compared. The familiar star-shaped urban structure is most evident in the first map, created at the end of 1996. From then onward, this star pattern gradually dissolves as fill-in development occupies desirable land around GZ. The result is that, in the final map, unclaimed land, shown as dark green, has largely vanished.

Studying the maps in detail reveals many other interesting features. For example, a complex lattice of roads, represented by the fine black lines, criss-crosses the plain. Areas of light blue, denoting rivers, ponds and lakes, also stand out. Another notable phenomenon is "sky writing", an artefact of the mapping process itself. Here, AlphaWorld citizens have deliberately claimed land in patterns that, when mapped, reveal words on the ground. Noticeable examples include the words "PLATTER" and "RAGU", visible in the map on page 207.

Perhaps the most striking single feature on the three maps is the large black area at the top-middle of each map. This is the result of an extremely large "land claiming" project that dates back to the earliest days of AlphaWorld, according to Vilett. Whilst the land is claimed, very few structures actually exist at this location, and yet it cannot be redeveloped for the reason explained above.

One must also be aware there is much that is not revealed in the maps. For example, because of the scale of the map, each pixel in the image has to represent a 10×10 meter plot on the ground; as a consequence, many small features are not shown. In addition, AlphaWorld is a three-dimensional environment, with buildings of different heights and some features located underground; these differentials are not captured by the two-dimensional maps (but see the 3-D map of a virtual world by Andy Hudson-Smith on page 210). The map also does not reveal how active and well-used different parts of the city are. Unfortunately, many areas of AlphaWorld are "virtual ghost towns", where land and buildings have been abandoned by their original homesteaders, but which cannot be reoccupied. And, because this is cyberspace, the buildings forever remain pristine clean, never decaying; the gardens never become overgrown; and no one ever visits!

4.21: Satellite maps of GZ City

chief cartographer: Roland Vilett (ActiveWorlds.com, Inc.).
aim: to show the detailed urban form of the major city growing at Ground Zero at several snapshots in time.
form: 2-D color raster map, looking like a satellite map of a real-world city with its messy, organic, urban morphology.
technique: generated from the AlphaWorld building database using custom software.
dates: December 1996, February 1998, and August 1999.
further information: see <http://www.activeworlds.com>

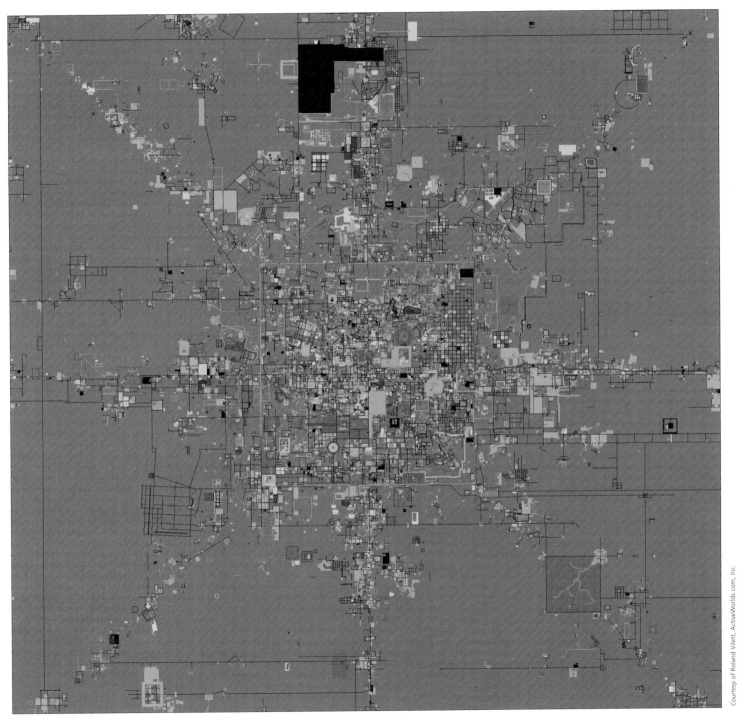

December 1996

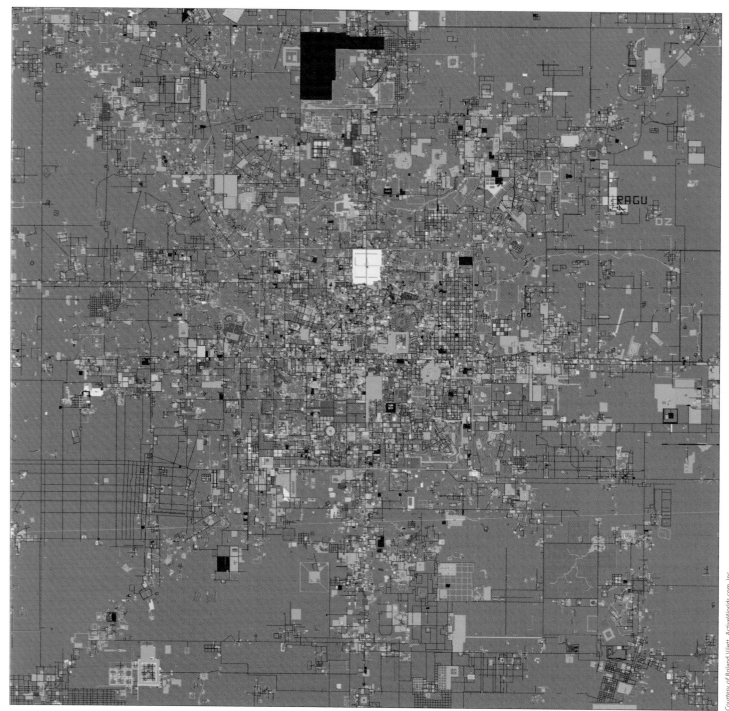

February 1998

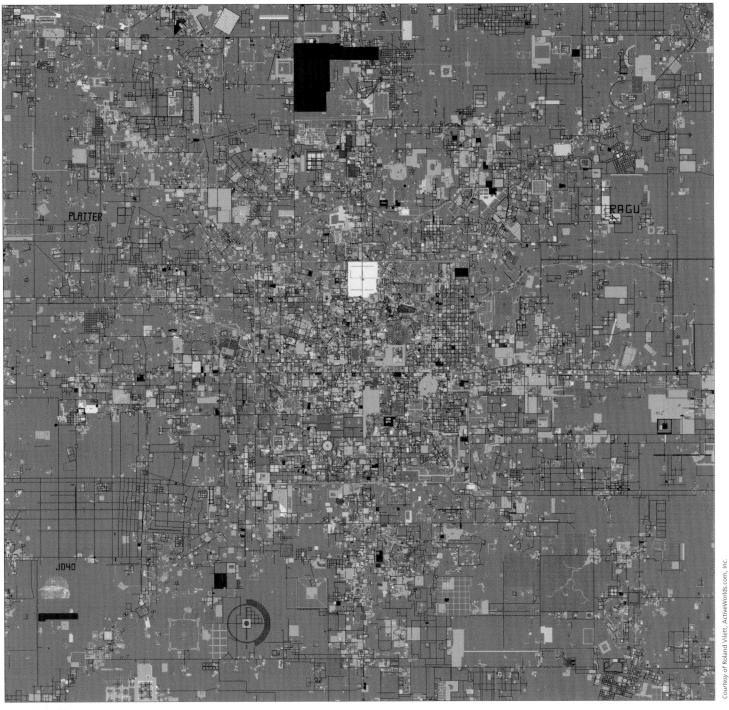

August 1999

Vilett's cartography provides us with city-wide "satellite" maps, but if we want to see more of the local detail and texture – what the buildings, trees and structures are really like – we need more detailed scale mapping. Such sophisticated mapping for AlphaWorld has been provided by Greg Roelofs and Pieter van der Meulen. Their system maps AlphaWorld at 12 different scales, and the images (clockwise from top left) show a zoom through ten of these. These scales are organized according to a pyramidal structure, with each layer being double the size and double the resolution of the one above. At the apex of the pyramid is the largest scale map, where the whole of AlphaWorld fits on a single screen image. Moving down through 12 layers of the pyramid, the base layer is a vast map where each screen image represents an area of 540 × 540 meters on the ground. The transition of scale from base to tip represents a zoom of over 2,000 percent. The maps can be browsed interactively, panning in all directions as well as moving through the different scales. Currently, all maps are updated every three months.

At the largest scale the map is mainly a green featureless expanse because urban development does not show up well, unlike in the density maps. As the scale increases, the urban detail becomes more visible. The most detailed scales are essentially equivalent to aerial photos, with each pixel in the large image representing one meter on the ground. This reveals the fine detail of roads, ponds, trees and buildings, along with localized "sky writing". The highest resolution map can actually be used as a powerful teleportation tool, placing the user in the landscape mapped.

4.22: Vevo multiscale mapping system

chief cartographers: Greg Roelofs and Pieter van der Meulen (Philips Silicon Valley Center, Sunnyvale, California).
aim: to map the whole of AlphaWorld at 12 different scales (from 20,000 to 500 feet).
form: color raster maps, where at the most detailed scale the maps appear aerial-photo-like, revealing individual buildings and roads.
technique: generated from the AlphaWorld building database using custom software.
date: images show the urban development of AlphaWorld, 5 November 2000. The system has been operational since July 1997.
further information: see <http://awmap.vevo.com>

655km × 655km

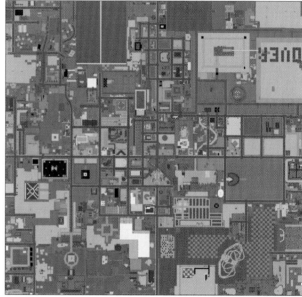

1km × 1km

2km × 2km

276km × 276km

138km × 138km

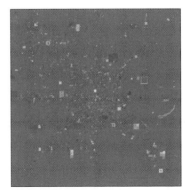

69km × 69km

ground zero

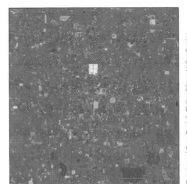

35km × 35km

4km × 4km

9km × 9km

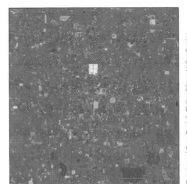

17km × 17km

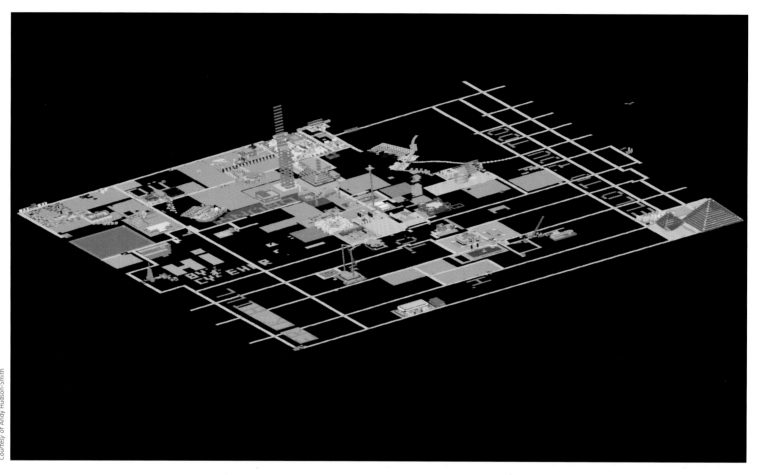

Courtesy of Andy Hudson-Smith

Using the same 3-D Active Worlds software platform as AlphaWorld, Andy Hudson-Smith has undertaken an interesting experiment in online community development. The experiment ran for 30 days initially, starting on 30 November 1998, and was a key element of his doctoral research concerning online planning. Hudson-Smith was particularly interested in the ability of inhabitants to design and build virtual architecture and communities. He therefore examined empirically the processes through which virtual place-making occurs. He created a new world called "30 Days" in which users were free to build a new community. The land extent of this world was 2 million square metres, and was capable of supporting 32 simultaneous users. Both registered and tourist users were able to enter the world and build structures, and no specific guidelines were provided as to where to build or what to build. A small prize was offered for the best structure built during the first 30 days of the project.

Hudson-Smith monitored in detail the building of urban structures and the social interaction of inhabitants. Maps of each day's urban development were produced and posted on the project's website. Most of these maps were two-dimensional, similar to those of Roland Vilett and the Vevo project which we have just examined. However, Hudson-Smith also created some innovative three-dimensional maps.

The example opposite shows the full extent of urban development in the "30 Days" world as of 4 January 1999. The map was produced from the building database for the virtual world using the Awmap mapping utility developed by user "Foxy". The map is displayed using an isometric projection, looking from the north-west corner of the world. Major structures projecting upward can clearly be discerned, the most noticeable being the purple tower in the middle – a kind of skyscraper building created by an unknown tourist – and two blue pyramids at the south-west corner of the world, created by user "Tom Hoxton". The structures are colored according to their owner. The map shows the building work of 89 different registered users, as well as an unknown number of tourists, who altogether placed nearly 29,000 objects. Interestingly, the map reveals several key geographic features of "30 Days" that are similar to AlphaWorld – for example, the regular grids formed by road builders (the yellow lines) and the sky writing.

4.23: 3-D isometric map of the urban environment of "30 Days"

chief cartographer: Andy Hudson-Smith (Centre for Advanced Spatial Analysis, University College London, UK).
aim: to visualize the urban environment developed in an online community experiment hosted by an Active Worlds virtual world.
form: 3-D isometric map showing built structures. Land-use colored by owner.
technique: map generated automatically from the building database of the virtual world, using the Awmap utility developed by user "Foxy".
date: 1999.
further information: see <http://www.onlineplanning.org>

In virtual worlds, the ability to teleport to distant locations is a vital form of movement. Andy Hudson-Smith produced an experimental 3-D model showing the location of fixed teleports that users had created in the "30 Days" world. Any object can be made into a teleport by giving it that "action" with specified destination coordinates. People then just touch the object with their avatar (representation) and they are transported to that location. There are two types of teleports, in-world and out-world. The in-world teleports are between two locations within the "30 Days" world, while out-world teleports link to locations into other virtual worlds in the Active Worlds universe.

Right are two rendered images of the model from different views. The "30 Days" world is shown in a transparent bubble, with the blue band representing the horizon. The teleports are shown by differently colored elliptical ribbons that pass through the origin and destination point. The flat plane in the middle of the bubble represents the extent of the land, and it has a 2-D map on it to provide context. In total, 65 teleports are shown, where 27 are in-world and the rest (38) are out-world. The teleport ellipses that touch the outer sphere are out-world teleports to other worlds. It should be noted that there may well be teleports in these other Active Worlds into the "30 Days" world, but these could not be determined from the data available to Hudson-Smith.

Analysis and mapping of the "30 Days" experiment is ongoing.

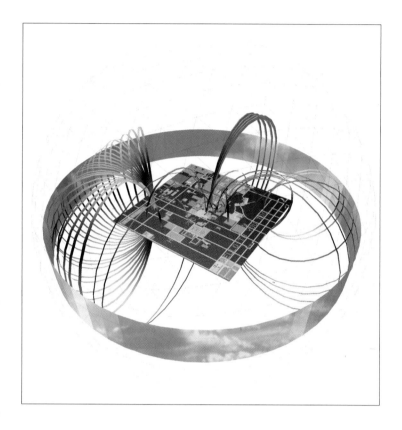

4.24: Teleports in "30 Days"

chief cartographer: Andy Hudson-Smith (Centre for Advanced Spatial Analysis, University College London, UK).
aim: to show the location and destination of teleports in the "30 Days" virtual world.
form: 3-D model with teleports shown as ellipses projecting through a 2-D map on the ground plane.
technique: locations of teleports and their destinations have been determined from the building database and then manually modeled in a CAD package.
date: 1999.
further information: see <http://www.onlineplanning.org>

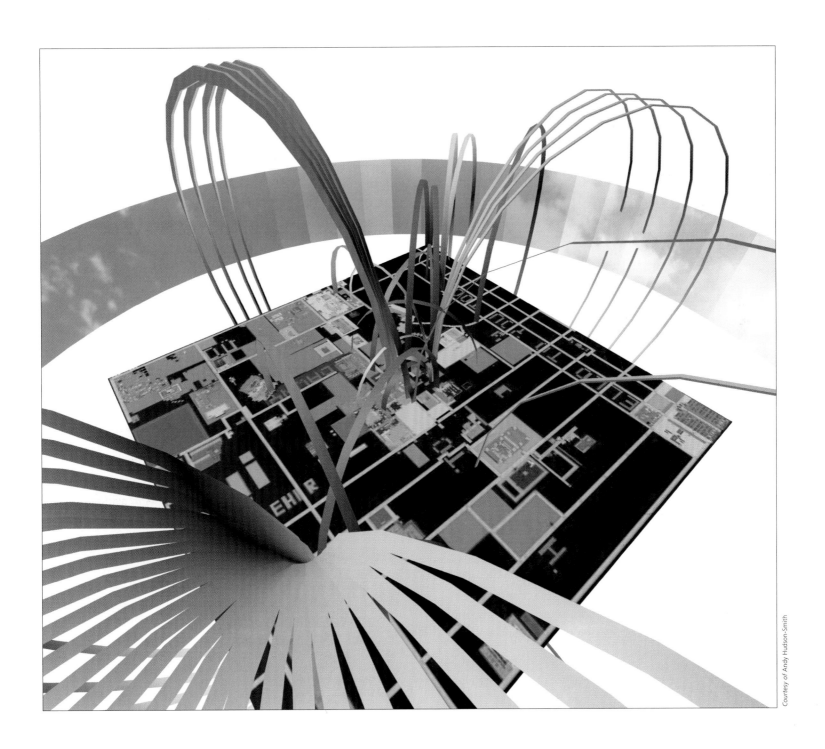

Mapping conversation and community　　**213**

Mapping game space

Computer games have been at the forefront of the development of virtual space. Since the release of pioneering games such as "Pong", "Pac-man" and "Donkey Kong" over two decades ago, the fierce pace of technical development has pushed games to offer ever more exciting and more immersive virtual experiences. Given the profitability of the computer games industry, recent trends have included using cyberspace as a distributed medium to increase market share. As such, the mass-market sales of powerful home PCs with 3-D graphics potential, coupled with the Internet, have given rise to networked games played by many thousands of players simultaneously.

The best of these games offer perhaps the most dynamic and interactive realms of cyberspace, although it is a realm often focussed on killing your opponent. But even in this, there is considerable social interaction despite it ending with a blast from a shotgun! The two major genres that work best as networked games are the first-person "shoot 'em ups", such as "Doom", and the so-called MMORPGs (Massively Multiplayer Online Role-playing Game), of which "EverQuest" and "Ultima Online" are the leading examples. These games are very popular – tens of thousands of people are immersed in cyberspace playing against one another. And this is their real power: the knowledge that the opponents are other people and are not computer simulated. The overall objective of many of the games may be to kill your opponent (often as quickly as possible) and yet it is still a vibrant form of social cyberspace, with communications and communities evolving. For example, teams are often forged in the guise of clans and guilds, and there is a rich fan culture associated with the games.

Before we start to review some of the maps of EverQuest, we detail in brief some of the characteristics of the game. The designers state:

Welcome to the world of EverQuest, a real 3-D massively multiplayer fantasy role-playing game. Prepare to enter an enormous virtual environment – an entire world with its own diverse species, economic systems, alliances, and politics. Choose from a variety of races and classes, and begin your quest in any number of cities or villages throughout several continents. From there, equip yourself for adventure, seek allies and knowledge, and head out into a rich world of dungeons, towers, crypts, evil abbeys – anything conceivable – even planes and realities beyond your imagination. Learn skills, earn experience, acquire treasure and equipment, meet friends and encounter enemies. A multitude of quests and adventures await, but you choose your role, you define your destiny. But whether you make yourself a noble human knight, a vicious dark elf thief, a greedy dwarven merchant, or whatever suits your desire, remember one thing: *You're in our world now.*

EverQuest is probably the leading massively multiplayer role-playing game on the Internet at present. As the designers note, it is a large three-dimensional virtual world in which tens of thousands of players can interact, go on adventures, and fight each other. The setting of the virtual world, the lands of Norrath, are very much the stuff of conventional role-playing games, consisting of a very Tolkienesque landscape. Unlike some of these games, EverQuest also includes money and a dynamic economy controlled by player actions. Individual adventuring is supported, but the game encourages the formation of parties to complete quests. Its environment is rendered in realistic three-dimensional graphics, from a first-person perspective. It has a large and varied geography, with many towns, villages, castles and ruins spread across several continents.

This playing environment is created by the game designers and remains fixed. This contrasts with more mutable virtual environments such as AlphaWorld or MOOs. The top map opposite is the stylized official game map. The screenshots below show various scenes and some of the characters that live there. There have been a number of notable efforts to map parts of the EverQuest game world. These are mainly hand-crafted maps created by the players themselves to help guide each other.

4.25: EverQuest

chief designers: developed by 989 Studios /Verant Interactive, released by Sony Online Entertainment.
date: March 1999.
further information: see <http://www.everquest.com>
further reading: These EverQuest community websites contain a huge wealth of Information: <http://eqvault.ign.com>
and <http://www.eqcorner.com>

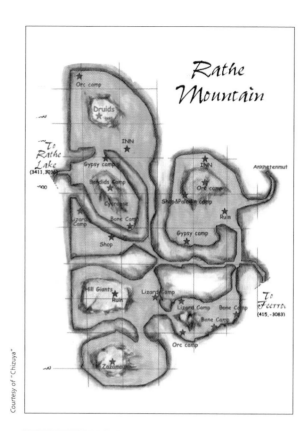

Rathe Mountain

Orc camp

Druids

INN

To Rathe Lake
(3411,3036)

Gypsy camp

Bandids Camp

Cyrenat

Lizard Camp

Bone Camp

Shop

INN

Orc camp

Shop&Paladin camp

Ruin

Gypsy camp

Ankhetenmut

Hill Giants

Ruin

Lizard Camp

Lizard Camp

Bone Camp

Bone Camp

Orc camp

Zazamoukh

To Feerrot
(415,-3083)

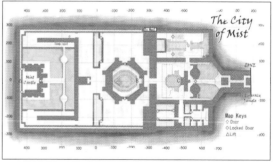

The City of Mist

Mist Castle

Map Keys
◇ Door
◇ Locked Door
△ Lift

Field of Bones

3823 / 0

N

Tangrin

Sythrax
2850 / -450

Beach

Maurders
1950 / 3400

Beach Cut
1750 / 1450

Cliff

Cliff

Beach

Warsliks Wood
1400 / 4300

Temple

Kurn's Tower
750 / 1300

0 / 3543

209 / 2514

Field of Bones

Bandit Camp
-560 / 3335

-296 / 2638

Empty

Dragon Skull

Field of Bones

Emerald Jungle
-1900 / -1260

Ancient City

-2480 / 3881

Kaesora
60 / -4190

East Cabilis
-2584 / 3548

Swamp of No Hope
-3400 / 1000

Gesler
Outriders of Karana
5/15/2000

The Map Shop of Norrath provides a comprehensive selection of more than 50 maps of the lands within EverQuest. The maps are created by player "Chizuya", based in Japan. They are high-quality hand-drawn maps, richly colored using a watercolor effect, containing detailed information on key points of interest. There are maps for major cities, wilderness areas – swamps, forests and mountains – as well as detailed layouts for castles and dungeons. They represent many hours of work researching and drawing.

We have chosen two examples that illustrate the quality of the maps available from Chizuya's Map Shop. The first (top-left) shows the Rathe Mountains, which lie in the east of the main continent of Antonica. The map shows grey and rocky mountain peaks, surrounded by lush green valleys, as well as significant locations such as camps, shops and inns. It is drawn on a regular coordinate grid so that a user can pinpoint locations and calculate approximate distances; it also shows connections to maps of adjoining areas. The second example (bottom-left) is a detailed layout of the rooms of the City of Mist, a small village in the Emerald Forest within the Kunark region. It shows individual rooms and whether their doors are locked or open. Again, the map is drawn on a regular grid.

Cartographer and member of the guild of the Outriders of Karana, "Gesler", produced a series of 12 maps showing in detail the geography of the Kunark region of EverQuest. (Kunark is an expansion of the original EverQuest world released in 1999.) The Field of Bones map delineates clearly the key features of the area with labeling and coordinates. The area is at the heart of the Kunark region and was once the site of a reservoir for the nearby city of Cabilis, which was the site of a great battle in the distant past. The area is largely desert and notable for the monsters (such as various types of skeletons, scorpions and scaled wolves) that the adventurer is likely to encounter. On 15 June 2000, the guild decided to disband.

4.26: The Map Shop of the Norrath

chief cartographer: player "Chizuya" (from Japan).
aim: to map portions of EverQuest.
form: colorful, watercolor-look map on a base grid. Many significant places are marked. Examples show the Rathe Mountains and the City of Mist.
technique: hand-created. More than 50 detailed maps of regions, cities and dungeons from the lands of Norrath.
date: 2000.
further information: see <http://www.nx.sakura.ne.jp/~chizuya/index2e.html>

4.27: The Kunark Mapping Project

chief cartographer: player "Gesler" (guild of Outriders of Karana).
aim: to produce a range of high-quality maps of Kunark region.
form: high-quality, two-dimensional schematic map.
technique: hand-created digital map.
date: May 2000.
further information: see <http://www.tapr.org/~OutridersKarana/index.html>

The EQ Atlas is the definitive source of maps and information on the geography of EverQuest. It was created and is maintained by player "Muse" (Michael A. Swiernik). The site is continually updated and represents a valuable community resource. More than 100 separate areas are covered by the EQ Atlas, with large areas covered by several maps. The maps themselves are well-designed and well-drawn sketch maps, shaded with a light pencil effect.

The example displayed opposite is an image of Ak'Anon, a city on the continent of Faydwer. The map is drawn with a coordinate grid and significant locations are numbered. (Detailed descriptions of the locations are listed on the map's accompanying Web page.) For example, location 5 is the "Abbey of Deep Musing – Cleric Trainers and Merchants selling Blunt Weapons, through secret door lies stairs down to Rogue Guild with Merchants selling Rogue Weapons". A large system of water channels and lakes flows throughout the city.

Ak'Anon has numerous guilds for clerics, necromancers, warriors and wizards. Two inset maps provide layouts of building complexes: the bottom one is a major shopping district and connects to the main map at point A; the top one is a maze of old caverns to the north of Ak'Anon that, according to the EQ Atlas, "have been taken over by the less desirable elements of the gnome society, where the necromancers and the other evil gnomes have taken up residence, as well as the undead that they have summoned to protect their home".

EQ Atlas provides valuable information and descriptions that accompany the map, giving details on the area's local character and its neighboring regions. The description for Ak'Anon begins: "An ancient city, built into the edge of the mountains long ago by the gnomes who call it home. Half fortress and half laboratory, the gnomes have been developing their clockwork machines and defenses here for generations, and show no signs of stopping."

4.28: The EQ Atlas

chief cartographer: player "Muse" (Michael A. Swiernik).
aim: to provide the most comprehensive atlas of EverQuest so far, with detailed maps of all areas of EverQuest along with useful descriptions of dangers and travel tips.
form: 2-D sketch map showing EverQuest environment.
technique: hand-constructed map.
date: September 2000. Atlas is continuously updated.
further information: <http://www.eqatlas.com>

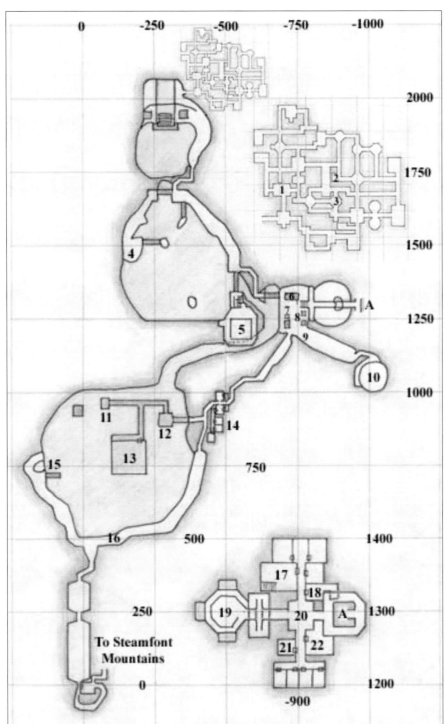

Ultima Online (UO) is another notable massively multiplayer online role-playing game. It has much in common with EverQuest, although it pre-dates it by a couple of years and consequently has less sophisticated graphics. Instead of a first-person perspective on the world, Ultima Online's interface is a third-person overhead isometric view. The screenshots (opposite-bottom) show the gaming environment. The current version of UO is based on a successful series of single-player role-playing games known as Ultima.

The geography of the UO virtual environment is the world of Britannia, consisting of nearly 200 million square feet, accompanied by subsequent expansions of The Second Age. Maps of every city can be used to guide the players. Like EverQuest, the land of Britannia has its own economic flux, political turmoil, battles and wars, and even a simple model ecosystem. Similarly, like EverQuest, actions have a real and lasting impact on the world of UO, and users run businesses and wage wars, although the terrain is fixed.

"Gram" has produced the best overview map (above right) of the Ultima Online world of Britannia so far. It took many hours of volunteer labor to create. It is a well-designed map, drawn on a regular grid, with stylistic elements to suggest the mythical realm of the game, such as a parchment effect and the style of lettering. Towns, dungeons and shrines are identified. The top corners of the map sheet are filled with supplementary information.

Another popular way to map the landscape of Britannia is to take multiple screen captures from within the game. Setting the viewpoint high above the terrain, a user can get a kind of aerial photograph view of the world. Moving across the world taking screen captures provides tiles that can be carefully stitched together to give a seamless map. The example shown top-opposite is the town of Yew (known as the City of Justice). Merchants and other points of interest are indicated, such as the Bloody Thumb Woodwork and the Court of Truth.

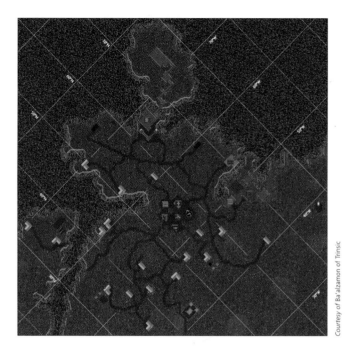

Courtesy of Ba'alzamon of Trinsic

4.29: Ultima Online

chief designer: Richard Garriott, Origin Systems, Inc., released by Electronic Arts.
date: September 1997.
further information: see <http://www.uo.com>
further reading: these leading Ultima Online community websites contain a huge wealth of information: <http://uovault.ign.com/> and <http://uo.stratics.com/>
"Killers have more fun", by Amy Jo Kim, Wired magazine, May 1998.
<http://www.wired.com/wired/archive/6.05/ultima.html>

4.30: Map of Britannia

chief cartographer: "Gram" (Graham P. Colwell).
aim: to provide a map of Britannia.
form: 2-D sketch map.
technique: hand-constructed map.
date: 1998.
further information: see <http://uo.stratics.com/homes/gramatls.shtml>

4.31: The Grand Atlas of Britannia

chief cartographer: Ba'alzamon of Trinsic (David G. Perkins).
aim: to produce maps of the major towns and dungeons.
form: somewhat like an aerial photograph, where one can discern roads and individual buildings.
technique: mosaic of multiple screen captures from the game.
date: 1999.
further information: see <http://uo.stratics.com/atlas/>

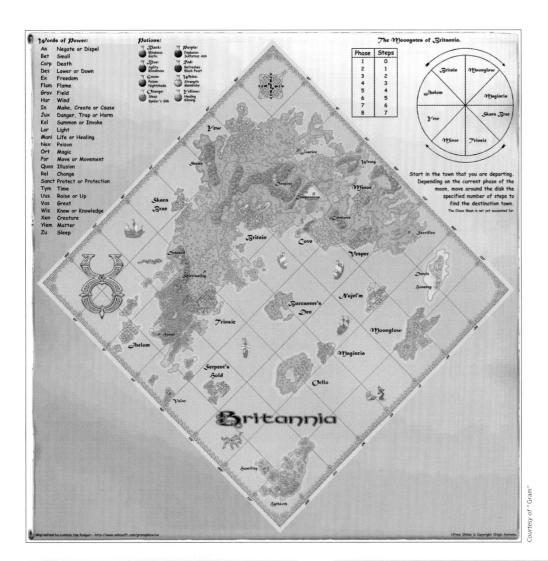

Courtesy of "Gram"

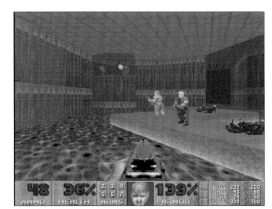

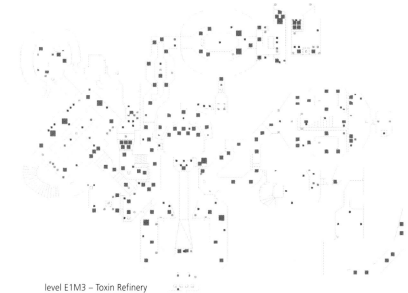

level E1M3 – Toxin Refinery

The computer game Doom, released by id Software at the end of 1993, pioneered a whole new genre of fast-action, violent "shoot 'em up" games. The key technical developments of Doom were its graphical rendering of the game world and its game play. In terms of graphics, Doom used a 3-D first-person perspective, where the game environment was viewed from a player's perspective, as can be seen in the screenshots opposite. This gives a very immersive feel to the gameplay, placing the player at the heart of the action. Doom, however, was not the first game to use first-person perspective; that honor probably belongs to the tank game Battlezone, released by Atari in 1980. Also, the immediate forerunner of Doom, Wolfenstein 3-D by id Software, developed many of the techniques used in Doom. Doom, though, became a smash hit – and the introduction to this view of virtual space for many millions of players.

The other major advance of Doom was its ability to have multiplayer games over local area networks and modems. In this violent realm of cyberspace, players had the thrill of knowing they were fighting against and killing human opponents, not just sterile computer-controlled "baddies". In multiplayer DOOM, the so-called DeathMatch mode, where every player is out for himself and himself (or herself!) alone, was particularly popular. So, even in this action-game space, it was still about interaction but mediated through gameplay (with rocket launchers!).

Doom also made extensive use of the Internet as an important channel to distribute the shareware version of the game to millions of players at very low cost. The actual software code of the game also had an open architecture that allowed players to modify, and experiment with, game design. Players could design new levels, new weapons and enemies. This greatly encouraged the development of a vibrant fan culture and enabled creative players to extend both the game environment and their enjoyment of it.

Opposite we present a number of screenshots of the first-person action from the Doom shareware game and two plan views of the level layouts. The shareware release had eight levels, starting with The Hanger (top-right). The plan views were created using a level editor called WadAuthor by John B. Williston. The red symbols in the map are different baddies, the blue are survival objects and the green ones are weapons and armour. The individual level environments are relatively small and compact, particularly compared with, say, AlphaWorld or EverQuest. This ensures that the action is fast and furious, for there is never long before there is another opponent to fight.

The creative company id Software developed the game further, releasing Doom II in October 1994. Then after two years' development the next generation of multiplayer 3-D first-person "shoot 'em up" was released by id Software, named Quake.

4.32: Doom

chief designers: John Carmack, Adrian Carmack, Tom Hall and John Romero (id Software).
aim: a fast, action game – which pioneered a whole new genre of computer games using a 3-D first-person-perspective "shoot 'em up" style. Players run around a large enclosed maze, killing anything in their way. Doom pioneered multiplayer games, which added a whole new dimension to game play. Hugely successful.
form: great graphics (for its time), coupled with considerable violence and gore.
date: released in December 1993.
further information: see <http://www.idsoftware.com>
further reading: the official Doom FAQ listing at
<http://www.gamers.org/docs/FAQ/doomfaq/>
The Doom community site at <http://www.doomworld.com/>

Quake developed from the gameplay of Doom, providing greatly enhanced graphical realism. The 3-D environment was richer in texture and dank dungeon menace, and the kinetic violence was rendered so fast that one could become dangerously immersed in it. The environments had a genuine three-dimensional feel to them. The multiplayer capabilities in Quake were also greatly enhanced, enabling many players to "frag" (i.e. fragment) each other in death matches over the Internet. Once again, an open software architecture facilitated great creativity and effort from the user community in producing "mods" for Quake – such as new maps, weapons and skins. This has greatly extended and enhanced the Quake world – all through volunteer activity.

Some scenes of violent struggles against "fearsome" baddies from a single-player game of Quake are shown opposite. The baddies include zombies who throw nasty lumps of flesh, chainsaw wielding ogres who throw devastating pipe bombs when within range, and other aggressive and unearthly fiends. The fights are fast and bloody. The actual 3-D environments are not much larger than in Doom, but they are much more three-dimensional, with bridges, holes, lifts and walkways (Doom tended to all be on one level). A 3-D wireframe view and a plan view of one of the early levels – The Necropolis – are shown below. These images are complex and were created using Qwdquake, a simple tool created by Gunnar Dahlström for viewing Quake maps.

4.33: Quake

chief designer: John Carmack (id Software).
aim: The son of Doom. A hugely successful first-person-perspective "shoot 'em up".
form: great 3-D graphics with murky dungeon-like levels and considerable violence and gore.
date: released 1996.
further information: <http://www.idsoftware.com>
further reading: <http://www.planetquake.com>

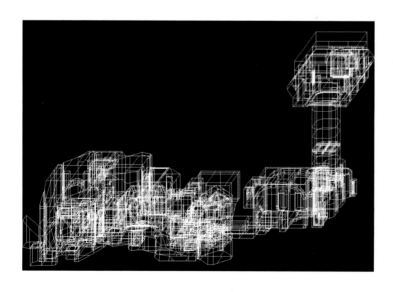

3D wireframe of level E1M3 – Necropolis

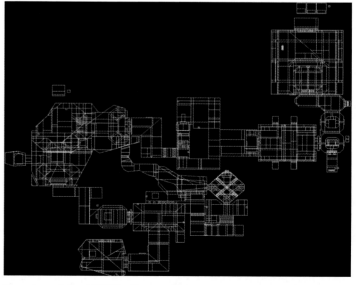

Plan view of level E1M3 – Necropolis

chapter 5

Imagining cyberspace

In this chapter, we turn our attention away from geographic and information visualization to consider four other ways in which cyberspace has been visualized and imagined spatially. The four alternative visions of cyberspace have been provided by writers, film makers, artists and architects. These four groups, like the researchers and software technicians we have so far discussed, have been seeking to answer the question "What does cyberspace look like?" These imaginary visualizations and mappings are important creative works in their own right, providing an often critical sphere in which to think about cyberspace and its structure, content and operation. They also have added relevance, however, because they often provide the inspiration and "blueprints" for designers and creators of the maps and spatializations discussed in previous chapters.

In other words they provide a popular imaginal environment in which to question and explore the space–time configuration of cyberspace – a cognitive space in which to think about the geographies of cyberspace. This, as we suggest below, has particularly been the case with fiction. As a consequence, although it might be easy to dismiss the visualizations we discuss in this chapter as "merely art", it should be recognized that this art often plays a wider role. As such, the influence of these architects, artists, film makers and writers should not be underestimated.

Science fiction visions of cyberspace

Science fiction (sci-fi) writing was undoubtedly an important genre throughout the twentieth century. Its importance lay in its ability to provide inspiration for those engaged in scientific enterprise and its function as a cognitive space in which to think about the consequences of technological development. Science fiction writing that has focussed on cyberspace has performed both of these roles. And just as cyberspace is a transformative technology, changing the way we live our lives, sci-fi that has considered cyberspace has been written within a new genre of science fiction writing – one that subverts its modernist traditions.

Cyberpunk recognized and explored our new post-modern condition through a literary vehicle that is itself decidedly post-modern. Here, we are not concerned with cyberpunk's exploration of how cyberspace will affect social relations, or how the style of cyberpunk challenged traditional models of sci-fi. Rather, we are concerned with its inspirational qualities and its role as a cognitive space, shaping the way that cyberspace has been conceived and developed.

In order to illustrate our point, we only consider the work of two writers, William Gibson and Neal Stephenson. These two novelists have been particularly influential in shaping the development, visual interface and spatial organization of cyberspace, and in articulating new geographic imaginations of emerging spaces such as the Internet. Indeed, it is now claimed by some that recent developments in both computing and society can be seen as an attempt to put their fictional visions into practice. This is not to say that other writers have not written about cyberspace and its space–time geometries or shaped the public imaginal sphere. Other influential cyberpunk fiction within this genre includes that by George Foy, Bruce Sterling and Tad Williams.

Gibson's first novel, *Neuromancer*, was a landmark book. Winner of the Hugo and Nebula awards, it is highly read and cited, in no small part due to the fact that it is the source of the word "cyberspace". It explored the possibilities of information and communication technologies long before the Internet was fashionable. All the subsequent novels, namely *Mona Lisa Overdrive* and *Count Zero*, which complete the "Sprawl" trilogy, *Virtual Light*, *Idoru*, and *All Tomorrow's Parties* describe a future world that has been reordered through libertarian capitalism and social Darwinism, and reshaped at all spatial scales through the socio-spatial processes of globalization and internationalization. As such, the global economy is dominated by a small number of transnational corporations, countries have fractured into weak nation-states, and society is divided even more than it is at present into the haves and have-nots, with the haves protected in gated communities and the poor left in relatively ungoverned, anarchic and lawless city suburbs. It is a time when knowledge is power, and information services, trade and espionage are major industries, with corporations and individuals linked by a vast computer network called "the Matrix".

Here, we are interested in Gibson's vision of the Matrix. His description of it (see boxed quotes) – a networked, Cartesian, visual, navigable dataspace – provided, it has been argued, the "imaginal public sphere" for computer scientists developing Internet and VR technologies. For example, in 1988 John Walker launched the Autodesk (a leading VR developer) "Cyberpunk Initiative". In a white paper entitled "Through the Looking Glass: Beyond User Interfaces", he invoked Gibson and proposed a project to produce a "doorway into cyberspace" within 16 months. As a consequence, many social scientists openly turn to Gibson to credit his foresight and acknowledge his influence in shaping the "Information Society". This is not to say that Gibson provided technical blueprints, but to acknowledge that he provided an initial cognitive space in which to think about cyberspace.

> Cyberspace. A consensual hallucination experienced daily by billions of legitimate operators, in every nation, by children being taught mathematical concepts … A graphical representation of data abstracted from the banks of every computer in the human system. Unthinkable complexity. Lines of light ranged in the nonspace of the mind, clusters and constellations of data. Like city lights, receding …
> *Neuromancer* (1984, p. 67)

> Program a map to display frequency data exchange, every thousand megabytes a single pixel on a very large screen. Manhattan and Atlanta burn solid white. Then they start to pulse, the rate of traffic threatening to overload your simulation. Your map is about to go nova. Cool it down. Up your scale. Each pixel a million megabytes. At a hundred million megabytes per second, you begin to make out certain blocks in midtown Manhattan, outlines of hundred-year-old parks ringing the old core of Atlanta.
> *Neuromancer* (1984, p. 57)

> . . . the infinite reaches of that space that wasn't space, mankind's unthinkable complex consensual hallucination, the matrix cyberspace, where the great corporate hotcores burned like neon novas, data so dense you suffered sensory overload if you tried to apprehend more than the merest outline.
> *Count Zero* (1986, p. 62)

5.1: Cyberspace in the "Sprawl" trilogy

writer: William Gibson.
books: "Sprawl" trilogy: *Neuromancer* (Ace Books, New York, 1984); *Count Zero* (Arbor House, New York, 1986); *Mona Lisa Overdrive* (Bantam, New York, 1988).
further information: William Gibson aleph by Manuel Derra; see <http://www.8op.com/gibson/>
further reading: *Mapping Cyberspace* by Martin Dodge and Rob Kitchin (Routledge, 2000). *Lost in Space: Geographies of Science Fiction*, edited by Rob Kitchin and James Kneale (Athlone Press, 2001).

They rose effortlessly through the grid, the data receding below them ... The fabric of the matrix seemed to shiver, directly in front of them... Somewhere far away, his hands moving over the unfamiliar keyboard configuration. He held them steady now, while a section of cyberspace blurred, grew milky.

Count Zero (1986, p. 231)

She spread the elastic headband and settled the trodes across her temples – one of the world's characteristic human gestures... She touched the power-stud and the bedroom vanished behind a colorless wall of sensory static. Her head filled with a torrent of white noise.

Her fingers found a random second stud and she was catapulted through the static-wall, into cluttered vastness, the notional void of cyberspace, the bright grid of the matrix ranged around her like an infinite cage.

Mona Lisa Overdrive (1988, p. 56)

A cubical holo-display blinked on above the projector: the neon gridlines of cyberspace, ranged with bright shapes, both simple and complex, that represented vast accumulations of stored data. "That's all your standard big shits. Corporations. Very much a fixed landscape, you might say. Sometimes one of 'em'll grow an annex, or you'll see a takeover and two of them will merge. But you aren't likely to see a new one, not on that scale. They start small and grow, merge with other small formations..." He reached out to touch another switch. "About four hours ago" – and a plain white vertical column appeared in the exact center of the display – "this popped up. Or in." The colored cubes, spheres and pyramids had rearranged themselves instantly to allow for the round white upright; it dwarfed them entirely, its upper end cut off smoothly by the vertical limit of the display... "and nobody knows what it is or who it belongs to."

Mona Lisa Overdrive (1988, pp. 254–55)

People jacked in so they could hustle. Put the trodes on and they were out there, all the data in the world stacked up like one big neon city, so you could cruise around and have a kind of grip of it, visually anyway, because if you didn't, it was too complicated, trying to find your way to a particular piece of data you needed.

Mona Lisa Overdrive (1988, p. 22)

He closed his eyes.

Found the ridged face of the power stud.

And in the bloodlit dark behind his eyes, silver phosphenes boiling in from the edge of space, hypnagogic images jerking past like film compiled from random frames. Symbols, figures, faces, a blurred, fragmented mandala of visual information.

Please, he prayed, *now* –

A gray disk, the color of Chiba sky.

Now –

Disk beginning to rotate, faster, becoming a sphere of paler gray.

Expanding –

And flowed, flowered for him, fluid neon origami trick, the unfolding of his distanceless home, his country, transparent 3-D chessboard extending to infinity. Inner eye opening to the stepped scarlet pyramid of the Eastern Seaboard Fission Authority burning beyond the green cubes of Mitsubishi Bank of America, and high and very far away he saw the spiral arms of military systems, forever beyond his reach.

Neuromancer (1984, pp. 68/69)

He is wearing shiny goggles that wrap halfway around his head; the bows of the goggles have little earphones that are plugged into his outer ears.

The earphones have some built-in noise cancellation features…

The goggles throw a light, smoky haze across his eyes and reflect a distorted wide-angle view of a brilliantly lit boulevard that stretches off into an infinite blackness. This boulevard does not really exist; it is a computer-rendered view of an imaginary place.

Snow Crash (1992, p. 20)

It is the Broadway, the Champs Elysées of the Metaverse. It is the brilliantly lit boulevard that can be seen, miniaturized and backward, reflected in the lenses of his goggles. It does not really exist. But right now, millions of people are walking up and down it.

The dimensions of the Street are fixed by a protocol, hammered out by the computer graphics ninja overlords of the Association for Computing Machinery's Global Multimedia Group. The Street seems to be a grand boulevard going all the way around the equator of a black sphere with a radius of a bit more than ten thousand kilometers.

Snow Crash (1992, p. 23)

The sky and the ground are black, like a computer screen that hasn't had anything drawn on it yet; it is always nighttime in the Metaverse, and the Street is always garish and brilliant, like Las Vegas freed from the constraints of physics and finance… If you go a couple of hundred kilometers in either direction, the development will taper down to almost nothing, just a thin chain of streetlights casting white pools on the black velvet ground. But Downtown is a dozen Manhattens, embroidered with neon and stacked on top of each other.

Snow Crash (1992, p. 26)

Like any place in Reality, the Street is subject to development. Developers can build their own small streets feeding off of the main one. They can build buildings, parks, signs, as well as things that do not exist in Reality, such as vast hovering overhead light shows and special neighborhoods where the rules of three-dimensional spacetime are ignored, and free-combat zones where people can go to hunt and kill each other.

The only difference is that since the Street does not really exist – it's just a computer-graphics protocol written down on a piece of paper somewhere – none of these things is being physically built. They are, rather, pieces of software, made available to the public over the worldwide fiber-optics network.

Snow Crash (1992, p. 24)

He is not seeing real people, of course… The people are pieces of software called avatars. They are the audiovisual bodies that people use to communicate with each other in the Metaverse.

Snow Crash (1992, pp. 35–6)

Neal Stephenson's novel *Snow Crash* was an international bestseller. As with Gibson's "Sprawl" trilogy, it charts a new and future information society. Although sharing some of the same concerns as Gibson (such as the long-term effects of libertarian capitalism), Stephenson envisages a different kind of geography – one of "burbclaves", a multispace city full of franchises. As with Gibson, information is a key commodity and the key resource is the Metaverse, a giant 3-D online community. Stephenson's main character, Hiro Protagonist, whilst a pizza delivery man in "real" space, is a main player (hacker) in the Metaverse. The story basically follows Hiro and his attempts to understand and combat Snow Crash – a virus within the Metaverse that flatlines its victims – and the corporation behind its use.

Not only did *Snow Crash* contain many visionary descriptions of an online world, the Metaverse, but it provided an obvious inspiration to many Internet and VR developers. Nowhere is this more evident than in the development of a 3-D virtual world, AlphaWorld, accessible across the Internet by the summer of 1995 (see chapter 4). AlphaWorld *is* a version of Stephenson's Metaverse, with the original world designers also adopting the names of the novel's main characters. Here, life is imitating art in a very literal sense. The significance of this translation from literature to reality should not be underestimated, especially when the product is as popular as AlphaWorld (which has over 800,000 unique visitors at the time of writing).

It is a piece of software called, simply, Earth. It is the user interface that CIC uses to keep track of every bit of spatial information that it owns – all the maps, weather data, architectural plans, and satellite surveillance stuff.

Snow Crash (1992, p. 106)

The room is filled with a three-dimensional constellation of hypercards, hanging weightlessly in the air. It looks like a high-speed photograph of a blizzard in progress. In some places, the hypercards are placed in precise geometric patterns, like atoms in a crystal. In other places, whole stacks of them are clumped together.

Snow Crash (1992, p. 214)

5.2: The Metaverse

writer: Neal Stephenson.
book: *Snow Crash* (Bantam, New York, 1992).
further information: Stephenson's homepage at <http://www.well.com/user/neal/>
also <http://www.sffworld.com/authors/s/stephenson_neal/>

Cinematic visions of cyberspace

Most films that have tried to emulate the success of the cyberpunk genre of sci-fi have largely failed to be big box-office draws. Many have also received a critical mauling. This in part has been, until very recently, because of technical limitations, so that even though the films are packed full of special effects they fail at some level of credibility. It has also been due to the poor quality of scripts. Typically, films within this genre have been driven by state-of-the-art special effects, with very limited narratives trying to hold the film together. It should also be noted that the post-modern tendencies of the literary genre fail to translate to the big screen, where they are portrayed within modernist frames of reference. As a consequence, films with multi-million-dollar budgets have failed to attract audiences. It is only relatively recently that the special effects that have been used so effectively in outer-space forms of science fiction since *Star Wars* (1977) have lived up to their full promise, allowing the potential of cyberspace's space–time geometries to be explored.

Despite the shortcomings, like its literary counterpart, sci-fi cinema has played an important part in shaping popular interest in cyberspace and virtual reality, and in providing an imaginal environment in which to contemplate the technology itself, its spatial formulation, and its wider social, political and economic implications. In this section, we detail three films in which cyberspace has been visualized, *Tron*, *The Matrix*, and *Warriors of the Net* (an online, non-Hollywood animation). Whilst the quality of the films is variable, their visualizations have shaped how we think about cyberspace. Other films that portray cyberspace include *Virtuosity, Hackers, Johnny Mnemonic, Lawnmower Man (2)*, and *eXistenZ*.

Tron pioneered the use of computer-generated special effects and also provided influential representations of virtual space within a computer. The plot charts the attempt by a video arcade owner, Kevin Flynn (played by Jeff Bridges), to hack into ENCOM, the company he used to work for and whose boss stole his video programes. His attempts at hacking are blocked by ENCOM's Master Control Program (MCP).

One night, after breaking into ENCOM yet again, Flynn is forcibly pulled into the virtual world by the MCP. In the virtual world, programs (who are the doubles of their users) are subjugated by the MCP, who tries to get the programs to pledge their allegiance to the MCP and renounce their users. If they refuse to do so, they are forced into gladiatorial contests in a Game Grid until they "de-rezz" (die). In order to defeat the MCP, Flynn must find and help TRON, a system security program. One of the most memorable scenes involves chases on light-bikes, as shown in the following images.

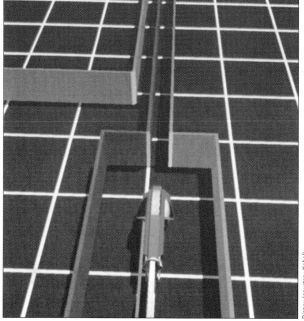

Disney (courtesy Kobal)

5.3: Tron
director: Steven Lisberger.
lead actors: Jeff Bridges and Bruce Boxleitner.
date: 1982.
further information: Guy Gordon's Tron page at <http://www.3gcs.com/tron/>

Imagining cyberspace

The Matrix, by Andy and Larry Wachowski, is a stunning action movie and was a major hit in 1999. The film charts the transition of Neo, from computer hacker to rebel warrior, on his discovery that life on Earth is nothing more than an elaborate façade – a false version of the twentieth century created to placate humans while their life essence is "farmed" to fuel the campaign of domination by the controlling "AI" in the "real" world (200 years in the future). Neo is contacted by Morpheus, who leads him into the real world and the fight against the Matrix. Neo is hailed as "The One", namely the person who will lead the humans to overthrow the machines and reclaim Earth. In order to overcome the Matrix, he has to battle his own doubts and also fight a series of "agents" used by the AI to fight the rebels.

The key representation of cyberspace in the film was the eerie green flowing computer code, culminating in the final battle with the AI agents in the "corridor of code". These striking images were created by Animal Logic, a special-effects company in Australia.

5.5: The Matrix

directors: Andy and Larry Wachowski.
lead actors: Keanu Reeves, Laurence Fishburne and Carrie-Anne Moss.
date: 1999.
further information: The Matrix website at <http://www.whatisthematrix.com>
further reading: The Art of the Matrix by Spencer Lamm (Newmarket Press, 2000).

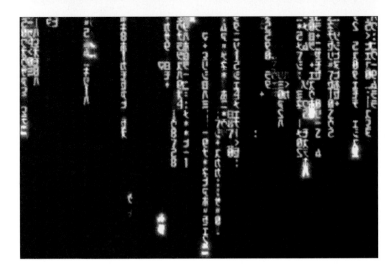

Explaining how the Internet works is a challenging task. How does that Web page get from some distant server onto your browser? How does the data move through the complex mesh of computers, servers and cables? A small team of artists at Ericsson Medialab in Sweden took on the challenge of explaining the inner workings of the Internet by using an animated movie. Entitled *Warriors of the Net*, the film follows an imaginary journey of data packets through different sections of the Internet (dispatch, local area network (LAN), routers, switches, firewalls, and undersea cables), accompanied by a narration. The inner workings of the Internet are represented as a dank Victorian world, with mechanical devices to move data packets physically on conveyor belts and up lifts. The images here are a chronological series of stills from the 14-minute movie.

5.6: Warriors of the Net

producer/director: Tomas Stephansson (Ericsson Medialab, Sweden).
animation: Gunilla Elam.
music/sound: Niklas Hanberger.
narration: Monte Reid.
aim: to show in a fun, non-technical way how the Internet works, by following an imaginary journey of data packets.
date: 1999.
further information: details and also a free download of the full movie at <http://www.warriorsofthe.net>

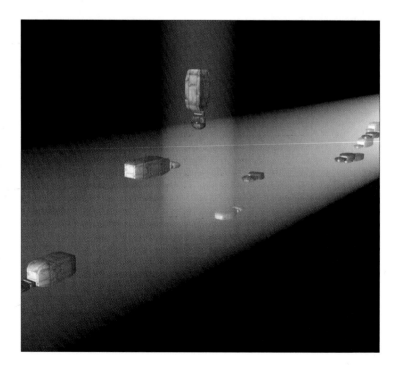

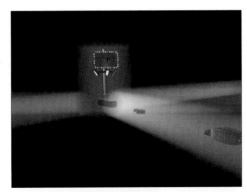

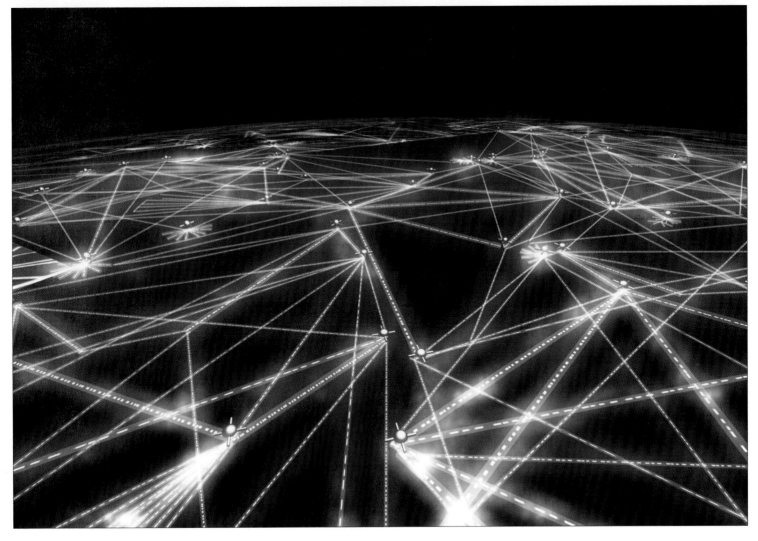

5.7: Shredder, RIOT and Digital Landfill

artist: Mark Napier (Potatoland.org).
aim: to subvert carefully designed Web pages displayed by conventional browsers.
form: collages of overlapping and jumbled images and text.
technique: created by algorithms from user-specified URLs, in a Java-powered Web interface.
date: Shredder: 1998; RIOT: 1999; Digital Landfill: 1998. Screenshots taken in September 2000.
further information: try out all Napier's art work at <http://www.potatoland.org>

Artistic imaginings: subversive surfing and warping the Web

Writers and film makers are not the only people to be exploring cyberspace's visual qualities. Artists, too, have been using cyberspace as a new medium of expression, and exploring its visual dimensions. Here, we are interested in the latter pursuit: how artists are portraying, visualizing and playing with cyberspace's form and extent. In particular, we focus on projects that either seek to subvert the Internet's conventions – especially how Web information is provided to us through browsers such as Netscape Navigator or Internet Explorer – or that seek to provide new ways of visualizing and mapping the Internet's complexity.

Mark Napier is a leading Net artist who has created an array of "subversive" browsing tools in his gallery/studio Potatoland. His work is unpredictable, disordered and unfinished, but it always provides interesting and interactive views of cyberspace. Here, we consider in brief three of his "anti-browser" projects – *Shredder, RIOT* and *Digital Landfill*. Each of these projects fractures the Web, providing raw and random collages of text and images, undermining thoroughly the design pretensions of those who carefully create Web pages. In particular, HTML code that is usually hidden is drawn into full view on screen, and images are stretched and piled upon each other in a chaotic jumble. Napier's projects recognize that, under the surface rendering presented in a browser, a Web page is a messy bunch of files, some of which are HTML code and scripts, and others are images and graphics. All of these files are sent as a stream of data that the browser software tries faithfully to put together according to the designers' wishes. Napier's work illustrates how these files can also be visualized in other ways, subverting their intended meaning and illustrating how information is created to promote certain messages.

Shredder was one of the first of Napier's art works and it simply shreds a given Web page into constituent parts. These are then "sprayed" onto the screen in a random order, size and position. The example top-left shows the shredded remains of

Andy Hudson-Smith's Online Planning homepage (http://www.onlineplanning.org). The hyperlinks are still clickable so that one can still try to browse as normal, but the design is rendered totally unintelligible. According to Napier on his website, *Shredder* "appropriates the data of the Web, transforming it into a parallel Web. Content becomes abstraction. Text becomes graphics. Information becomes art."

RIOT is a more recent and sophisticated project, producing collages of Web randomness. However, rather than concentrate on a single page, *RIOT* deconstructs images and text from different Web pages and throws them all together. Moreover, it is also multi-user, and so the resulting collage can be seen, shared and extended by other users accessing *RIOT* at the same time across the Internet. Napier claims that "*RIOT* dissolves the territorial boundaries on the Web". Example screenshots far left, middle and bottom, and top-right show the effects *RIOT* has when turned loose on Web pages of the Queen Mother, cats and the author's homepages (kitchin.org, cybergeography.org) respectively.

The final Napier art project is called *Digital Landfill*. It is another collaborative, multi-user work where users can dump their old, unwanted data trash and thereby create a dynamic work of art. The contents of the eponymous digital landfill, and therefore the work of art itself, change constantly as new layers of digital waste accrete. As Napier says: "The artwork, like the Web, is a cumulative group effort. It takes as input all the 'stuff' that netizens throw into it. Ultimately the viewers decide what the landfill will look like." On initiating access to *Digital Landfill* you can choose to add some new waste or go and view the current state of the landfill. If you choose to add some of your own trash, this can be in the form of unwanted email, text, HTML code, or image URLs. When you view the landfill, you see a slice of it, as shown in the bottom-right screenshot on page 240. The listing of layers down the left-hand side shows the date and title, and these allow you to move up and down through the landfill, viewing different layers.

One aspiration of Net artists has been to warp the accepted norms of Web presentation by subverting the carefully crafted designs of Web content. The results show a new, unseen, and often disordered view of cyberspace. One of the most notable of these "subversive" anti-browsers is called Web Stalker. The presentation is very different from how most users see the Web. For a start, Web Stalker does not render any pictures or graphics. It was developed by a three-person art- and-design collective based in London in 1997. It is minimalist software, stripped down of functionality so that it can be distributed on a single diskette.

On starting Web Stalker, a user is presented with a single, large, black window that fills the screen. The window is completely blank, with no hint of how to use the browser or what it offers in terms of functionality. Onto this empty canvas the user can draw boxes, which are then assigned a certain action or function. The software only provides six functions: Crawler, Map, Stash, Extract, Stream and Dismantle. Function boxes can be of any size and position on the canvas, and they can overlap each other.

For us, the most interesting aspect of Web Stalker is the Map function, on which the creators say "the mapping is dynamic – 'Map' is a verb rather than a noun". Map draws the hyperlink structure surrounding a given URL as delicate circles and lines. These white filaments build up over time to form views of small areas of the Web, similar in appearance to the output from a Spirograph. The images right show snapshots of different Maps of several websites. It is quite hypnotic to watch the images gradually form over several minutes as the Web Stalker crawls through the hyperlinks. They are not designed as practical tools for Web management or navigation (as, say, the examples we surveyed in chapter 3), but they act somewhat as abstract X-rays of the hidden structures of the Web.

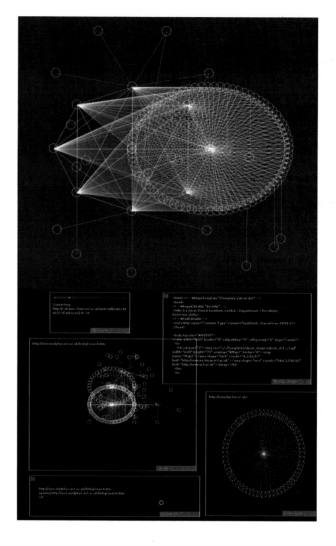

5.8: I/O/D4 – The Web Stalker

artists: Matthew Fuller, Simon Pope, Colin Green (I/O/D, London).
aim: to provide an anti-browser that subverts the conventional page representations of the Web.
form: a blank, black canvas onto which various function boxes can be drawn, including a Spirograph-like Map function.
technique: specialized browser software.
date: 1997. Screenshots taken in September 2000.
further information: see <http://bak.spc.org/iod/>
further reading: "The Browser is Dead" by Arie Altena, *Mediamatic*, 9(4), October 1999; <http://www.mediamatic.nl/magazine/9_4/altena_browser/altena_1gb.html>
"*A Means of Mutation*", by Matthew Fuller, March 1998, at <http://bak.spc.org/iod/mutation.html>.

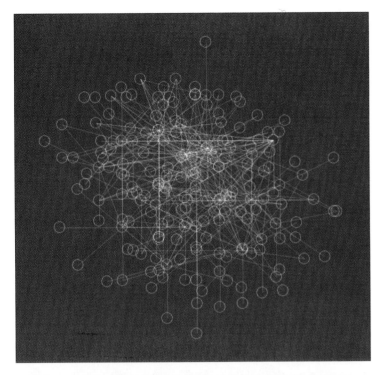
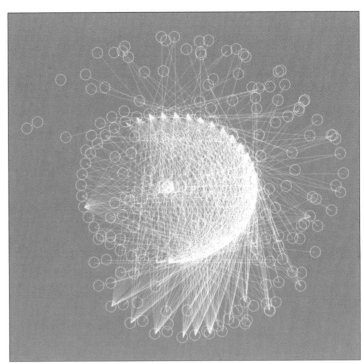
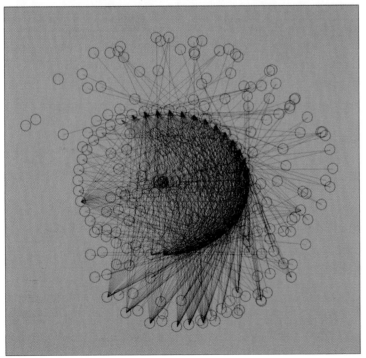
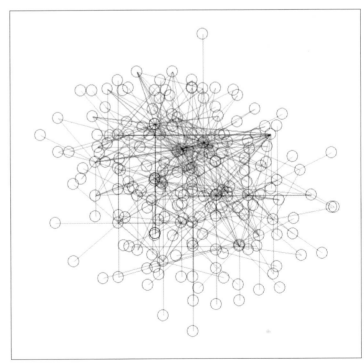

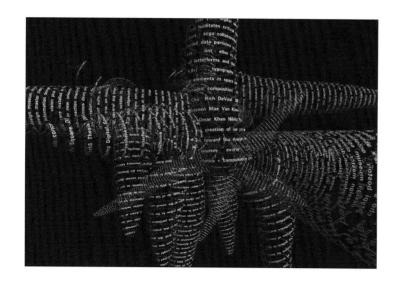

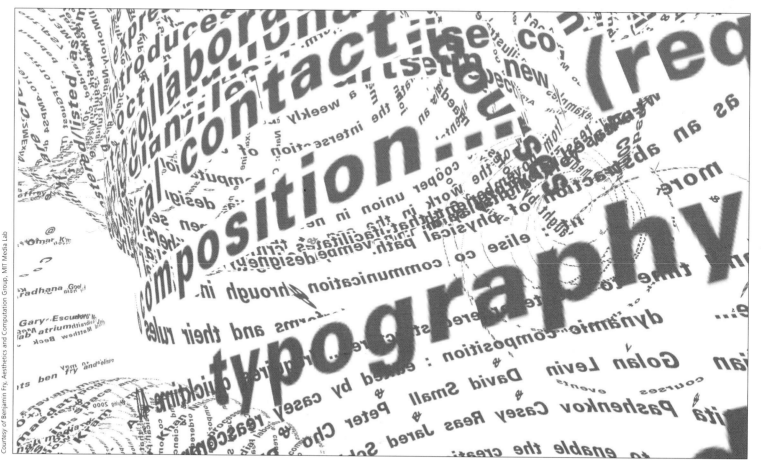

The images here show sculptured, structured tendrils of text representing, in an abstract manner, the content of a website. They were created by Ben Fry, whose work on organic information visualization we discuss more fully in chapter 3.

In this instance, his aims are art-like rather than precise information visualization. To create the sculptures, Tendril first reads a Web page and analyzes where that page links to. A branch or tendril-like structure is made from the text content of the Web page, and the same is done for each of the linked pages, with the linked branches attaching themselves to the main system. Over time, the result is a huge branching structure, built from the text contained in a set of connected Web pages.

5.9: Tendril: typographic sculptures from Web content

artist: Ben Fry (Aesthetics & Computation Group, Media Lab, MIT).
aim: to construct sculptures from Web content.
form: spiraling tendril-like sculptures of text.
technique: custom-written code.
date: 2000.
further information: see <http://acg.media.mit.edu/people/fry/tendril/>

Pinpointing where things are on the Web is determined by universal resource locators. These define precise locations, and they are used for navigation via hyperlinks. Links are the foundation of the Web. Linkie, the "Link Machine", is a visual directory of links that presents them as a random scatter. However, Linkie does not provide any extra information or judgement. Simple white links fill the screen in a chaotic, overlapping barrage of potential connections and then slowly dissolve, to be replaced with a new spatial configuration. Linkie displays a user-entered database of links from which it randomly selects one to display.

The credits for Linkie say that "we found the link machine inside a crackerjack box" and it is very much a playful interface to the Web. It is one of the works of D+CON/trol, a semi-monthly online gallery of new media, whose mission is reported as "an experimental exercise in the loss of viewer control over the developing Web medium". The idea is "to manipulate the viewer response to content that is normally provided through linked visual guideposts to information". In other words, exploring the Internet through Linkie is designed to show the ways in which how we browse and search cyberspace are guided and shaped by media designers.

Much of our browsing and searching of the Web is directed by the major portals and search engines such as Yahoo!, AltaVista, Lycos and AOL. However, these portals only provide a partial and subjective mapping of the Web, and media artist Andy Deck has sought to confront these issues in his CultureMap work. He describes CultureMap on his website as 'a visualization of proportion, disproportion, direction, and indirection in the content and no content of the World Wide Web'. Like Linkie, the project is designed to make people think about *how* they find information and *what* they find. Deck is a media artist based in New York who has been creating Web-based art since the mid-1990s. He focusses on collaborative drawing spaces, game-like search engines, problematic interfaces and informative art.

CultureMap is an abstract visualization that highlights the daily volume of "hits" that certain portals (for instance altavista.com,

alltheweb.com, google.com) report for 32 categories of information, as defined by broad generic keywords such as "shopping", "travel", "music", etc. Users select which categories they are interested in, and these are visualized as brightly colored tiles. The size of the tile in the resulting map is scaled to reflect that day's average reported "hits". A user can explore the map by clicking on the tiles to bring up an interactive dialog box offering various functions. These include Fluctuation and Hit Tracker, which let the user see in detail how the volume of "hits" changes over time as either an animated tile map or stock-ticker style of display.

Deck says that CultureMap is "joining abstract design with representation of the information landscape . . . [It] depicts the evolution of Web content in the form of a dynamic composition." This composition has much in common with some of the information maps we examine in chapter 3, especially ET-Map. However, CultureMap is not designed for practical navigation; it is instead a critical appraisal of information categorization on the Web. As Deck concludes, "Any mapping of the Web's content is bound to be imprecise. CultureMap does not pretend to reveal any rigorously objective picture of the content of the Web."

5.10: Linkie – Link Machine (v2)

artists: D+CON/trol – Andre F. Sousa and Devon Bleak (OneTenDesign).
aim: to show a changing directory of user-submitted Web links.
form: a random collage of plain white URLs that pulsate and slowly redraw every few seconds.
date: 2000. Screenshot taken in October 2000.
further information: see <http://www.decontrol.com/>

5.11: CultureMap

artist: Andy Deck (Artcontext.org), commissioned and displayed by Turbulence.
aim: to show the structure of the Web by measuring the amount of information in different categories, as measured by major search engines.
form: colored tiles representing the different categories, with size related to the volume of information.
technique: automatic, daily sampling of 32 different categories (like shopping, travel or music) from search engines, using keywords.
date: 2000. Screenshot taken in October 2000.
further information: see <http://www.turbulence.org/Works/cultmap/>
see also Andy Deck's homepage at <http://artcontext.com>

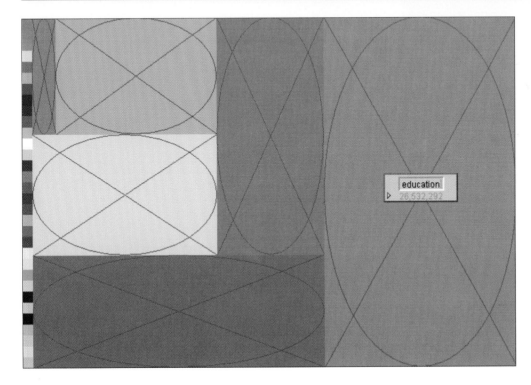

The artists of this work claim that the Natalie Bookchin and Alexei Shulgin Universal Page is "the objective average of all content on the Web merged together as one" and "a pulsating, living monument commemorating no single individual or ideology but, instead, celebrating the global collective known as the World Wide Web". It looks to all intents and purposes like a page of randomness, a gobbledegook sprawl of text. However, it is derived from scripts that crawl the Web and employ "precise algorithms" to generate "average content". The system behind the Universal Page works continuously to keep it updated in real time.

5.12: The Universal Page

artists: Natalie Bookchin and Alexei Shulgin; programming by Alexander Nikolaev.
aim: to provide an objective summary of the whole Web as a single page
form: a screen of jumbled text and spaces.
technique: claims to be based on "precise algorithms".
date: 2000. Screenshot taken in October 2000.
further information: see <http://www.universalpage.org>

[every: access]

This dense mosaic of multi-colored pixels is reminiscent of an image of interference on a TV set. It provides a snapshot of one dimension of the whole Web on a single screen, where the variable pattern of horizontal striations indicates the different densities of IP address allocation and usage across the total range of possible addresses. The "every" interface is also an image map, so that each pixel is a hotlink to the website that it represents. So it is both a map of, and an interface to, the Web. In conclusion Jevbratt says (again on his website) of the 1:1 project interfaces that they "are not maps of the Web but *are*, in some sense, the Web. They are super-realistic and yet function in ways images could not function in any other environment or time. They are a new kind of image of the Web, and they are a new kind of image."

The 1:1 project tries to visualize the structure of the Web that is usually hidden from view. It is the structure of IP addresses (looking like, for instance, 200.93.167.214) that computers on the Internet use to route data to each other. Jevbratt's project reveals this structure by providing five different interfaces to a continuously monitored database of IP addresses of all the Web servers on the Internet. This database was initially developed for another art project at C5 and is maintained automatically using softbots (agent software) that trawl the Internet. The title of the project – 1:1 – makes reference to the famous fictional allegories of mapping at the one-to-one scale by Lewis Carroll and Jorge Luis Borges. Jevbratt on his website says that, using 1:1, "first we encounter a collapse between the map and the interface. But the postphotographic practice of the 1:1 project makes the implosion even more severe. The interface becomes not only the map, but the environment itself".

The five interfaces of 1:1 are "random", "excursion", "petri", "hierarchical" and "every". The above image shows a screenshot of the "every" interface. This provides a visualization of every IP address of the Web as a single image. Each IP is represented by a pixel in the image, where the color is determined by the last three numbers in the IP address.

5.13: 1:1 – "*every*" interface

artist: Lisa Jevbratt (C5 and The CADRE Institute, San Jose State University, USA).
aim: to map Internet protocol (IP) addresses of the Web at a scale of 1:1 as a new form of interface. This forms the C5 IP database used in five different 1:1 mappings.
form: "every" is one of five different interfaces. It shows every IP address of the Web as single pixels in a large image. The color of each pixel is based on the individual numbers in the IP address.
technique: Softbots search the total IP address space to identify Web server addresses.
date: 1999. Screenshot taken in October 2000.
further information: see <http://c5corp.com/1to1/index.html>

The Electric Sky map is a piece of art, depicting a constellation motif, that acts as an image map to link a range of different Net art works. As well as providing links to sites, the map also reveals academic and cultural liaisons that existed in 1996. These links were easier to discern then than now, because only a few museums, galleries and other repositories for art had registered their own domain names; instead, many collaborated with a university or with commercial servers, forming nested art aggregates. The map depicts such collaborative networks as constellations, with the primary hub of the network in red. Solid blue lines indicate direct collaborations, while dotted lines indicate indirect links. Jon Ippolito, the artist/curator, states on his website: "As the nighttime sky offered ancient mariners a readymade navigational chart, so The Electric Sky . . . offers modern-day voyagers a map with which to steer their way across the World Wide Web." Electric Sky is now archived as part of an interactive gallery of artistic Web maps known as the CyberAtlas, curated at the Guggenheim Museum by Jon Ippolito himself.

5.14: Electric Sky

artist: Jon Ippolito assisted by Danny Piderman (CyberAtlas, Guggenheim Museum).
aim: to provide an interactive map of Net art sites.
form: star-chart, where different classes of art site have different styles of star symbol.
technique: simple interactive 2-D image map.
date: spring 1996.
further information: see <http://www.guggenheim.org/cyberatlas/home/index.html>

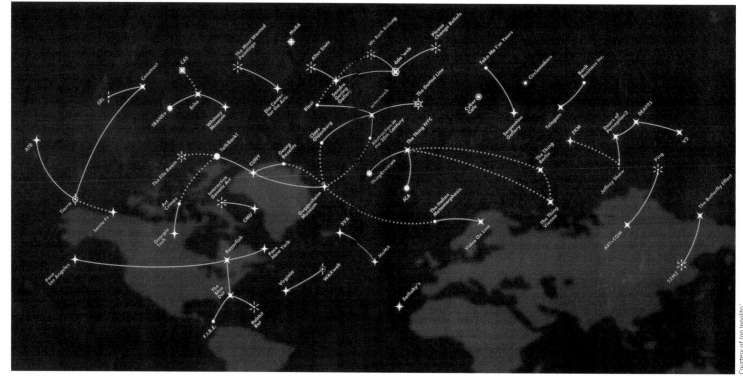

Imagining the architecture of cyberspace

Like writers, film makers and artists, a number of pioneering architects have begun to explore the actual and potential architecture of cyberspace. In this section we discuss the work of two projects that seek to delimit, and project through new means, the spatial form of cyberspace's architecture. This work is important because it challenges visualizers of cyberspace to extend their analysis beyond conventional understandings of space and to explore new ways of thinking about how cyberspace might be visually conceived. In both cases, complex algorithms are used to "compose" new architectural forms.

In his 1991 article, Marcos Novak, founding director of the Laboratory for Immersive Environments and the Advanced Design Research Program at the School of Architecture, University of Texas, Austin, argues that cyberspace has a "liquid architecture":

Liquid architecture is an architecture that breathes, pulses, leaps as one form and lands as another. Liquid architecture is an architecture whose form is contingent on the interests of the beholder; it is an architecture that opens to welcome me and closes to defend me; it is an architecture without doors and hallways, where the next room is always where I need it to be and what I need it to be. Liquid architecture makes liquid cities, cities that change at the shift of a value, where visitors with different backgrounds see different landmarks, where neighborhoods vary with ideas held in common, and evolve as the ideas mature or dissolve.

"Liquid Architectures in Cyberspace" (1991, pp. 251–2)

Here, Novak is arguing that cyberspace has a spatial and architectural form that is dematerialized, dynamic and devoid of the laws of physics; spaces in which the mind can explore free of the body; spaces that are in every way socially constructed, produced and abstract. As such, the architecture of cyberspace only mirrors that of Cartesian logic if that is how we straitjacket it. In his work, over several projects, he has been trying to redefine how we think about cyberspace's architecture, seeking to push our understanding beyond its Cartesian logic – beyond our own lack of imagination.

In other words, like the artists we have discussed earlier in this chapter, Novak is seeking to subvert how we imagine cyberspace's form and spatial structure. In order to do this, he has produced a whole series of images that seek to articulate cyberspace's "liquid" form. Some of these images are displayed as on pages 252–53 and represent some of his more recent work, entitled ie4D. The images are created using algorithms to "compose" architecture and are four orthogonal views of the same space.

5.15: ie4D

architect: Marcos Novak (visiting professor at the Department of Architecture and Urban design, UCLA).
aim: to explore the "liquid architecture" of cyberspace.
form: mutable landscapes.
technique: "grown" by genetic algorithms.
date: 2000.
further information: Novak's website at <http://www.centrifuge.org/marcos/>
further reading: "Liquid Architectures in Cyberspace" by Marcos Novak, in *Cyberspace: First Steps*, 1991, edited by Michael Benedikt (MIT Press), pp. 225–254.
"Trans Terra Form: Liquid Architectures and the Loss of Inscription" by Marcos Novak, at <http://www.t0.or.at/~krcf/nlonline/nonMarcos.html>

GINGA is an acronym for Global Information Network as Genomorphic Architecture, and this is a project designed to examine the spatial representation of information in cyberspace. It is a 3-D browsing system based on a large collection of digital information, developed by Japanese architect Fumio Matsumoto. Using special algorithmic codes, Web resources are reconfigured by GINGA into one of nine three-dimensional worlds: Nebula, Ring, Network, Forest, Strata, Text, Image, Polyphony, and Cemetery (see opposite) respectively. In Nebula, for example (top-left), information is distributed according to its data identifier, such as its URL or IP address, and those sites that share similar common URLs/IP addresses are clustered together. In Ring, (left, second from top), information is assembled into ring structures, which are grouped according to the type of information. The ring's diameter is determined by the frequency of update and its width by the volume of data. In Network (large picture), the linkage between nodes is displayed, with the size of link reflecting traffic flow. In Forest (left, third down), a tree directory structure is represented, with sharp-pointed trees containing nested pyramids of directories where their height represents the levels of hierarchy. In Strata (middle column), the chronological order of information is represented. In Text (middle-right), an archive of text information is presented as thin layers. In Image (bottom-left), pictures, maps and photos are arranged in a random floating pattern to create a "labyrinth of memory". In Polyphony (bottom-middle), sound is converted into visible bits that circle columns representing individual artists, instruments or music types. In Cemetery (bottom-right), dead avatars are stored in an arranged order.

Users can explore each of these nine worlds using avatars, examining how different kinds of information might be most effectively visualized in cyberspace.

5.16: GINGA – nine worlds of cyberspace

architect: Fumio Matsumoto with Shohei Matsukawa (Plannet Architectures, Tokyo).
aim: to examine how different kinds of information architecture might be most effectively visualized in cyberspace.
form: nine different 3-D "worlds", each employing a different mode of information representation.
technique: VRML modeling and an interactive interface to explore the different kinds of information architecture.
date: 2000.
further information: see <http://www.plannet-arch.com/>

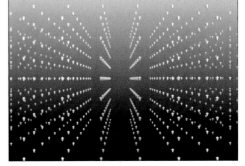

Imagining cyberspace   **255**

chapter 6

Final thoughts

The atlas of cyberspace that we have presented here is wide-ranging but is inevitably a subjective sample of publicly available maps at a particular point in time. Although we kept adding new examples up until our December 2000 deadline, we are sure that several new and innovative mapping methods will have been developed by the time this book is published, and we encourage you to seek out other maps of cyberspace. Moreover, while we have provided some commentary, we would suggest three supplementary activities.

The first is to actually try out some of the spatializations. Using the software is in many ways the only way of gaining a full comprehension of the sophistication of the maps or spatializations developed and appreciating how they work. This is particularly the case for immersive and dynamic spatializations, which are not best represented on the printed page. Many of the examples we detailed are freely available on the Web or to download. Good starting places include VisualRoute (page 65), NewsMaps (page 118), Chat Circles (page 174), and Map.net's 3-D cityscape (page 147).

The second activity is to investigate the story behind a map, following the links within the text to explore researchers' and company websites and the articles they have written about their work. Often, the story behind the map is as interesting as the map itself. In order to investigate the maps and spatializations, and to keep up with new developments, we suggest that you visit <http://www.cybergeography.org>, an evolving repository of articles and examples of maps and spatializations, with up-to-date links to appropriate websites.

The third activity is to use the discussion detailed in chapter 1 to question the images displayed and to think through ways in which the maps and spatializations might be improved – and why so many of our examples, although interesting visually, fail in practicality. Because of this last point, we have sometimes been asked whether the maps and spatializations are, and will continue to be, nothing but 'eye-candy' – nice to look at but of little practical or analytical use. Although some maps are undoubtedly little more than this (which might be

expected, given the prototype status of many), a number of these views of cyberspace improve our understanding of it and others hold great potential. For example, although there are problems with ecological fallacy (see page 5) in relation to maps of infrastructure, the maps still reveal important information about the extent and capacity of different networks, and they also provide valuable insights into how social and spatial relations are being transformed. Such spatializations, even though having limited use at present, provide a useful experimental basis from which more practical methods can be developed and from which widespread application is likely to emerge.

Given the debate about 'eye-candy' or usefulness, we think – based on our research – that there are a number of ways that a project of mapping cyberspace can be advanced. They are:

- to more fully explore real-time mapping, using dynamically generated data through measuring the network itself;

- to develop sophisticated spatially-referenced and temporal data sources;

- to ensure these data are standardized;

- to produce informative meta-data about underlying data;

- to explore ways of determining the relative accuracy of spatializations and to acknowledge and display potential errors;

- to improve spatial legibility – that is, to establish how easy the map is to interpret (drawing on cartographic theory and cognitive science);

- to test the usability of spatializations, using those data to update mapping techniques;

- to study the effects of spatialization on the media mapped;

- to establish collaborative links between disparate groups working on related themes (and we hope that this book might help in the process);

- to appreciate and account for ethical and privacy implications of mapping cyberspace;

- to examine the effects, on the maps themselves, of the social context within which mapping takes place;

- to extend the coverage of maps and spatializations to include media that have so far received little attention, such as email; to include any new media developed; and to account for current trends, such as mobile access devices, broadband access, the continuing diffusion of technologies across and within societies, and the end of English-language dominance on the Web.

Although there is a long way to go, we are confident that the utility and power of mapping cyberspace will be revealed in the coming years.

And remember: there is no one *true* map of cyberspace.

Further reading

A small selection of useful and interesting books and articles is listed below for those who want to investigate further.

Abbate, J. (1999) *Inventing the Internet*. MIT Press, Cambridge, Mass.

Anders, P. (1998) *Envisioning Cyberspace: Designing 3D Electronic Space*. McGraw-Hill, NY.

Benedikt, M. (1991) *Cyberspace: first steps*. MIT Press, Cambridge, Mass.

Berners-Lee, T. (1999) *Weaving the Web: The Original Design and Ultimate Destiny of the World Wide Web by Its Inventor*. HarperBusiness, New York.

Burgoyne, P. and Faber, L. (1999) *Browser 2.0: The Internet Design Project*. Lawrence King Publishing, London.

Card, S. K., Mackinlay, J. D. and Shneiderman, B. (eds) (1999) *Readings in Information Visualization: Using Vision to Think*. Morgan Kaufmann Publishers, San Francisco.

Damer, B. (1997) *Avatars! Exploring and Building Virtual Worlds on the Internet*. Peachpit Press, San Francisco.

Dodge, M. and Kitchin, R. (2000) *Mapping Cyberspace*. Routledge, London.

Dodge, M. and Kitchin, R.M. (2000) 'Exposing the "second text" in maps of the Network Society', *Journal of Computer Mediated Communication* 5(4). http://www.ascusc.org/jcmc/vol5/issue4/dodge_kitchin.htm>

Hafner, K. and Lyons, M. (1996) *Where Wizards Stay up Late: The Origins of the Internet*. Simon and Schuster, New York.

Harley, J. B. (1989) 'Deconstructing the map', *Cartographica*, 26, pp. 1–20.

Harpold, T. (1999) 'Dark continents: critique of Internet metageographies', *Postmodern Culture*, 9(2), January. Also at <http://www.lcc.gatech.edu/~harpold/papers/dark_continents/index.html>

Herz, J. C. (1997) *Joystick Nation*. Abacus, London.

Janelle, D. and Hodge, D. (eds) (1999) *Accessibility in the Information Age*. Springer-Verlag, Berlin.

Kahn, P. (2000) *Mapping Websites: Designing Digital Media*. Rotovision, London.

Kitchin, R. (1998) *Cyberspace: The World in the Wires*. John Wiley and Sons, Chichester, England.

Kitchin, R. and Kneale, J. (eds) (2001) *Lost in Space: Geographies of Science Fiction*. Athlone Press, London.

MacEachren, A. M. (1995) *How Maps Work: Representation, Visualization, and Design*. Guildford, New York.

Monmonier, M. (1991) *How to Lie with Maps*. University of Chicago Press, Chicago.

Poole, S. (2000) *Trigger Happy: The inner life of videogames*. Fourth Estate, London.

Rheingold, H. (1993) *The Virtual Community: Homesteading on the Electronic Frontier*. Addison-Wesley, New York.

Salus, P. H. (1995) *Casting the Net: From Arpanet to Internet and beyond…* Addison-Wesley, New York.

Shields, R. (ed.) (1996) *Cultures of Internet: Virtual Spaces, Real Histories and Living Bodies*. Sage, London.

Smith, M. A. and Kollock, P. (eds) (1999) *Communities in Cyberspace*. Routledge, London.

Standage, T, (1998) *The Victorian Internet: The Remarkable Story of the Telegraph and the Nineteenth Century's Online Pioneers*. Weidenfeld & Nicolson, London.

Turkle, S. (1995) *Life on the Screen: Identity in the Age of the Internet*. Simon & Schuster, New York.

Wood, D. (1993) *The Power of Maps*. Routledge, London.

Index